Governing Marine Protected Areas

In this innovative volume, the author addresses some important challenges related to the effective and equitable governance of marine protected areas (MPAs). These challenges are explored through a study of 20 MPA case studies from around the world. A novel governance analysis framework is employed to address some key questions: how can top-down and bottom-up approaches to MPA governance be combined? What does this mean, in reality, in different contexts? How can we develop and implement governance approaches that are both effective in achieving conservation objectives and equitable in fairly sharing associated costs and benefits?

The author explores the many issues that these questions raise, as well as exploring options for addressing them. A key theme is that MPA governance needs to combine people, state and market approaches, rather than being based on one approach and its related ideals. Building on a critique of the governance analysis framework developed for common-pool resources, the author puts forward a more holistic and less prescriptive framework for deconstructing and analysing the governance of MPAs. This interdisciplinary analysis is aimed at supporting the development of MPA governance approaches that build social-ecological resilience through both institutional and biological diversity. It will also make a significant contribution to wider debates on natural resource governance, as it poses some critical questions for contemporary approaches to related research and offers an alternative theoretical and empirical approach.

Peter Jones is a Senior Lecturer in the Department of Geography, University College London (UCL). He has spent more than 20 years undertaking interdisciplinary and applied research on the governance of human uses of marine resources, particularly in relation to marine protected areas (MPAs) and marine spatial planning (MSP). He has provided advice to many national and international organisations on MPA and MSP issues, including the United Nations Environment Programme (UNEP), the Convention on Biological Diversity, the European Commission and the International Union for the Conservation of Nature (IUCN). He is also a Ministerial Appointee to the S   nservation Authority. He enjoys life by the sea   children.

'There are many books on how to design and create marine protected areas (MPAs), so vital for ocean conservation, but few on how to make them successful. This book plugs that crucial gap, distilling experience from across the world into sound and creative advice.'

Callum Roberts, Professor of Marine Conservation, University of York, UK

'The issue of governance – of how we manage the places and things over which we have responsibility, including the habitats and creatures of the sea – is addressed in this book in a novel and important way. Through the lens of his review of a large number of MPA case studies, Jones makes the case for diversity and complexity in the governance of MPAs, which are themselves ecologically diverse and complex. This is a signal achievement which should greatly advance both policy and practice.'

Bonnie J. McCay, Distinguished Professor, School of Environmental and Biological Sciences, Rutgers the State University of New Jersey, USA

'There is a lot of scattered information about the value of MPAs – indeed we do need them to protect marine biodiversity, and the more there are, and the bigger they are, the better it is. But we need a book that makes sense of all this information, informs us on how to make the best of the MPAs we have, and reviews why we need more. Dr Jones' is that book.'

Daniel Pauly, Professor of Fisheries, University of British Columbia, Canada

'MPAs are mystifyingly prone to failure. Using a robust case study analysis approach, Jones clearly demonstrates a broad variety of ways to achieve management-to-scale via different governance arrangements, in order to harness the significant potential of MPAs in achieving conservation and sustainable use objectives.'

Tundi Agardy, Executive Director, Sound Seas,
Author of Ocean Zoning: Making Marine Management More Effective
(Earthscan, 2010)

'Peter Jones examines a very timely topic: governing marine MPAs within nations' waters. He sees the big picture and the diversity of details that governments and stakeholders need to shape if MPAs are to benefit biodiversity and people. Read this book!'

Elliott Norse, Founder and Chief Scientist,
Marine Conservation Institute, Seattle, USA

'Here is a book that deals with governance of a natural system for purposes of resilience. It is really worth reading!'

C.S. Holling, Arthur R. Marshall Jr. Chair in Ecological Sciences,
Department of Zoology, University of Florida, USA

Governing Marine Protected Areas

Resilience through diversity

Peter J. S. Jones

Routledge
Taylor & Francis Group

LONDON AND NEW YORK

First published 2014 by Routledge

2 Park Square, Milton Park, Abingdon, Oxon OX14 4RN
711 Third Avenue, New York, NY 10017, USA

Routledge is an imprint of the Taylor & Francis Group, an informa business

First issued in paperback 2016

British Library Cataloguing in Publication Data
A catalogue record for this book is available from the British Library

Library of Congress Cataloging in Publication Data
Jones, Peter J.S.
Governing marine protected areas : resilience through diversity /
Peter J.S. Jones.
pages cm. -- (Earthscan oceans)
Includes bibliographical references and index.
1. Marine parks and reserves--Government policy. 2. Marine
biodiversity conservation--Government policy. 3. Environmental
policy. 4. Environmental protection. I. Title.
QH91.75.J66 2014
333.95'616--dc23
2013033368

ISBN 978-1-138-67923-8 (pbk)
ISBN 978-1-84407-663-5 (hbk)
ISBN 978-0-203-12629-5 (ebk)

Typeset in Goudy
by Saxon Graphics Ltd, Derby

To Jo, for being, providing and enabling
everything that is good about my life

Contents

List of illustrations

Figures

Tables

Boxes

Foreword

Although the seas, and particularly shallow nearshore waters, may seem familiar, many of the biophysical processes that drive life and ecosystem relationships within them differ from those of life on land. Most island and coastal societies have deeprooted cultural concepts of the seas as providers of food and processors of wastes, and of distant oceans that are vast and dangerous, but that offer inexhaustible riches and opportunities for the bold. Against this background, cultural acceptance that human activities can change and are changing the ecosystem processes of our seas is both recent and challenging.

As creatures of the land we manage land resources from legal and regulatory approaches that are deeply rooted in the definition of areas in relation to the surface of the land – drawing lines and defining boundaries. We assign rights and responsibilities on the basis that different uses and forms of management can occur either side of a boundary or fence, with limited cross-boundary interaction. These concepts do not transfer easily to marine environments.

On land there is unidirectional surface connectivity through waterflows within catchments from hilltop to rivermouth. Terrestrial conditions are severe for living cells but this is our cultural normality as mammals with physiologies evolved to enable cells to survive, thrive and reproduce, despite desiccation and climatic dynamics and diurnal temperature changes as great as 20° C. Life is sustained by primary productivity on the surface related to ecological processes in the top few metres of soil. The third dimension of the atmosphere sustains no biomass, but provides limited biological connection through spores, pollens, aerial seeds and animals that can fly or drift. In this context, biodiversity management through title and primacy of use enables areas of threatened habitats to be managed as islands of naturalness, provided catchment waterflow and wildlife linkage corridors are also addressed.

The normality of marine systems is that most primary production occurs in the upper water column. Many species develop from egg to adult form in the water column and during that process they are part of a complex food web. The entire life cycle of many species is spent in the water column, drifting in the plankton or swimming within or between water masses. The water column is also the nursery and distribution mechanism for the larvae of species that live on or within the seabed or have adult territories associated with seabed habitats and features, such as rocky or coral reefs. Such larvae may travel hundreds of miles before settling. Life in the lower water column and on the seabed is almost entirely reliant on secondary production, fed by energy from primary production many hundreds and even thousands of metres above. The seas are truly more complex and three-dimensional than the land to which we are accustomed.

Some island and coastal communities have maintained traditional marine management approaches which protect fishing areas and culturally important species, but for larger areas the drivers, multi-sectoral demands and long distance linkages in three-dimensional marine systems present major governance challenges. Terrestrial concepts based on primacies of use and sectoral governance within two-dimensional site or area based management of critical, and often relict, areas of natural habitat can be effective in the management of specific benthic site-dependent communities and species without a planktonic phase that brood or nurture their young. Thus, coral reefs and other specialised seabed communities, such as seamounts and submarine vents, can be managed on such a narrower spatial basis. More generally, however, the relationship between extensive sedimentary and rocky seabeds and the overlying water column creates a challenge in identifying specific areas that are more (or less) ecologically significant, so protected areas and conservation require an integrated multi-sectoral approach when extended out to the seas.

Nevertheless, as technologies extend the range and scale of human uses of marine space, no-take marine protected areas (MPAs) have increasingly essential roles in biodiversity conservation and the management of fisheries, along with other human uses and impacts. They are essential as control, reference and refuge areas, for informed scientific management of the relationship between humanity and the seas. Effective broader concepts of MPAs, including those subject only to partial protection from particular activities, are also needed for integrated management, ecosystem-based use and trans-sectoral engagement if we are to address the inextricably linked and most challenging issues of conserving ecosystem processes and marine biodiversity, and maintaining food and resource security.

Against this background, this book addresses the challenges of implementing and effectively managing MPAs for narrower habitat protection. It also places MPAs in the broader context in terms of both governance and ecological theories. These issues are addressed in a novel, thorough and inter-disciplinary way, reflecting the broader trans-disciplinary contexts of conservation within the marine spatial management of fisheries, water quality and the impacts of climate change through ocean warming, acidification and changing coastal dynamics.

The use of case studies illustrates the challenges, opportunities and constraints of different approaches to establishing and implementing effective MPAs. The range of case studies enables the reader to develop an understanding of the trans-disciplinary issues that have to be addressed in an integrated and effective policy framework. This is an important contribution to the field of marine conservation and natural resource management.

Richard Kenchington, Professor, Ecosystem and Resource Management,
Australian National Centre for Ocean Resources and Security,
University of Wollongong.

Acknowledgements

First and foremost, I must thank the United Nations Environment Programme (UNEP) for funding much of the research on which this book is based, particularly Ole Vestegaard of the Marine, Coastal & Freshwater Ecosystems Branch of the Division for Environmental Policy Implementation at UNEP, for his recognition of the potential of this research and subsequent support for it. I am also grateful to Carl Lundin (IUCN), for his initial suggestion to produce a guide to governing MPAs, to Dan Laffoley (IUCN), for facilitating the use of the World Commission on Protected Areas – Marine Network to recruit some of the case studies, and to Peter Mackelworth and his colleagues at Blue World for hosting and providing funding support for the MPA case study workshop. Many of the insights related to MSP have been gained through my involvement in the 'monitoring & evaluation of spatially managed marine areas' (MESMA) research project, and I am grateful to the European Commission for funding this project and to the many researchers around Europe who have collaborated in it.

The research on which this book is based has been inspired by discussions with too many people to name here, but I am particularly grateful to Wanfei 'Feifei' Qiu and Elizabeth De Santo, whose assistance and insights in their PhD research and as subsequent collaborators on this research have been invaluable, to other PhD students who have guided me into and through various theories and related literatures, particularly Matthew Fortnam, Peter Mackelworth, Tom Roberts and Minsuk Jun, and to Professor Jon Side, Heriot-Watt University, for suggesting and supervising my PhD research on MPAs, which has continued ever since. This research could not have happened without the enthusiastic input of the MPA case study contributors named and cited in Table 14, to whom I am extremely grateful. Thanks are also due to Miles Irving and Cath D'Alton in the Drawing Office, Department of Geography, University College London, for producing and redrawing the figures in this book, and to the UNEP-World Conservation Monitoring Centre, the American Association for the Advancement of Science, Elsevier and Macmillan Publishers for their permission to reproduce and adapt some of the figures in this book. Last but not least, I am grateful to my parents, Patricia (in memoriam) and John Jones, for always being there and encouraging me, and to my wife, Jo, and children, Kirsty and Nathan, for their emotional support and constant inspiration.

Chapter 1

An introduction to Marine Protected Areas

Marine Protected Areas as a governance challenge

Marine Protected Areas (MPAs) have become firmly established in theory, but much less firmly established in reality. There has been a rapid growth in recent decades in the number of papers and books that discuss the need for MPAs and the science of designing and monitoring them, but the rate at which MPAs are actually designated and effectively protected has not kept pace. Whilst 12.7 per cent of the global land area is covered by terrestrial protected areas, only 2.3 per cent of the global marine area (including high seas) and 5.7 per cent of the marine area under national jurisdiction (excluding high seas) is covered by MPAs,[1] despite growing concerns about declines in marine ecosystems and fisheries.

Why are those concerned about the degradation of our seas continuing to swim against the tide in their efforts to put effective MPAs in place? How can we move to MPA governance systems that balance the vested interests of marine industries, particularly fishing, with the growing scientific and societal concerns about the state of our seas? If it is accepted that MPAs are needed, who should have a say in where they should be and what activities should be excluded? How can the participation of local users in managing MPAs be balanced with the need to achieve strategic conservation objectives and fulfil related obligations? How can MPA restrictions be enforced and cooperation with them otherwise promoted? How can we ensure that the fishermen and related workers who lose out in the short term as a result of MPA restrictions are given a fair deal?

This book will explore these questions and the many issues that they raise from a governance perspective. Governance is discussed in terms of how the 'top-down' role of the state (including laws and regulations), the 'bottom-up' role of the people (including public and user participation) and the 'neoliberal' role of markets (including economic incentives and property rights) can be combined to promote the achievement of MPA objectives. Debates about how to better steer the management of MPAs will be considered in the wider context of debates about governance, with a particular focus on how different governance approaches, i.e. the role of states, people and markets, can be combined to bridge the gap between MPA theories and realities. During the 1990s, the focus of publications was on why

MPAs are necessary to restore marine ecosystems and promote sustainable fisheries exploitation, including selection criteria for individual MPAs and the rationale for representative networks. During the 2000s the focus shifted to the scientific basis of designing ecologically coherent MPA networks.

This book will build on such work, undertaken mainly by natural scientists, by focusing on how MPAs can be effectively and equitably governed, contributing to the increasing role that social science is playing in MPA debates and initiatives. These discussions will be grounded in the realities of governing MPAs, drawing mainly on 20 case studies from around the world, undertaken with funding from UNEP (Jones et al., 2011), in which governance issues have been systematically analysed employing a research framework developed by the author.

This book and the case studies it draws on are focused on seas under national jurisdiction,[1] i.e. the focus will not be on MPAs in high seas beyond national jurisdiction as these are subject to different international regulatory regimes which pose very particular governance challenges related to international laws and policies under the auspices of the United Nations and various regional fisheries management organisations. Whilst seas under national jurisdiction represent 39 per cent of the global marine,[1] they tend to support a higher intensity and diversity of uses compared to high seas, are relatively heterogeneous and diverse, and support relatively high levels of ecosystem productivity, the majority (80–90 per cent) of the world's fish catches being within seas under national jurisdiction.

This chapter will go on to outline growing concerns about declines in fish stocks and marine ecosystems, the resultant calls and targets for MPAs, their international policy framework, their history and the slow progress with their designation and effective implementation. Chapter 2 will discuss the main objectives of MPAs and relate these objectives to different categories of MPAs. Recognising that MPAs represent a progression and extension of their terrestrial equivalents, Chapter 3 will consider some key differences and divergences that often underlie governance challenges, particularly between different value priorities, between the views of marine ecologists and fisheries scientists, and between marine and terrestrial ecosystems. Chapter 4 will introduce the theoretical framework for the research on which this book is based, through a discussion of different perspectives on the governance of natural resources, protected areas and MPAs. Chapter 5 will describe the empirical framework that was developed and applied to explore the realities of MPA governance in case studies of MPAs from around the world. Chapter 6 will introduce the case study MPAs and overview the main findings of the governance analyses. Chapter 7 will focus on how the 36 incentives on which these analyses are based have been individually applied amongst the 20 case studies. Chapter 8 will focus on an analysis of how the incentives have been combined to constitute an effective and equitable governance framework, followed by an overview of the implications of this study for governance theories, for research on MPAs and for MPA practitioners.

A basic premise of this book is that whilst much research has been undertaken on MPA governance issues, advice on best practice essentially revolves around the statement that 'design and management of MPAs must be both top-down and bottom-up' (Kelleher, 1999, xiii). This book will focus on exploring what this actually means in practice, particularly the challenges of combining top-down, bottom-up and market approaches to governing MPAs, given that 'MPA governance' is the various processes by which decisions are taken and implemented, underlying what is technically described as 'MPA management'. It will be of interest to actual MPA managers in providing a balanced analysis of MPA governance challenges and options for addressing them, particularly in terms of the 36 incentives, which could serve as a menu for options for improving governance, including case studies of how they have been combined in different MPA contexts. It will also contribute to academic debates on natural resource governance issues, including in relation to the concept of social-ecological resilience, by challenging the dominant view that specific governance approaches, particularly bottom-up approaches, are 'best' or 'right'. This challenge will not, however, lead to calls to go 'back to the barriers' (Hutton et al., 2005) through a return to imposed, unjust command-and-control 'fortress conservation'. Rather, the emphasis will be on the need to *combine* governance approaches, recognising that different approaches each have their strengths and weaknesses.

The dichotomy between arguments for bottom-up and top-down approaches to protected area governance, along with a growing focus on market approaches, suggests that these approaches may be mutually exclusive. This book will argue and demonstrate that this is not the case, and that in the same way that having a higher diversity of species leads to ecosystem stability and resilience, using a higher diversity of governance approaches in combination leads to institutional stability and resilience.

Societal concerns extend out to sea

Some recent assessments of the state of marine ecosystems provide major causes for concern, but what can be done to address these concerns and improve the prospects? This section reviews these concerns and their related societal expressions, leading to an overview of calls for MPAs as a key means of addressing them.

Trends in the state of marine ecosystems in the last 25 years led the Millennium Ecosystem Assessment to conclude that 'most services derived from marine and coastal ecosystems are being degraded and used unsustainably and therefore are deteriorating faster than other ecosystems' (UNEP, 2006, 4). The assessment also concluded that 'arresting the further degradation of coastal and marine ecosystem resources for the benefit of both present and future generations is an urgent imperative' (UNEP, 2006, 6). Another report concluded that in the next 20 years fishing and pollution will increase, warning that 'the need to plan and implement ecosystem scale and ecosystem-based management of the seas is urgent' if society is to avoid the substantial deterioration of marine biodiversity and the 'growing consequences for [the] resource and physical security of coastal nations' (UNEP, 2010, 5–6). A subsequent report from the International Program on the State of the Oceans (IPSO, 2011, 3) similarly warned that there has been a 'serial decline' in the health of ocean ecosystems, due to multiple stressors, such as overfishing (direct and indirect impacts), pollution (toxics and eutrophication), habitat loss through coastal development, invasive species and climate change (ocean warming and acidification), and that 'without significant changes in the policies that influence human interactions with the marine environment, the current rate of ecosystem change and collapse will accelerate and direct consequences will be felt by all societies'. A recent book provides a compelling account of the fascinating nature of marine ecosystems but also of the growing threats to their health which could undermine their potential to deliver numerous critically important ecosystem services on which the human race depends (Roberts, 2012a), such as food provision, nutrient recycling, climate regulation and shoreline protection (Costanza et al., 1997; Beaumont et al., 2007).

In order to better assess the state of the world's marine ecosystems, a marine 'Dow Jones index' has been developed to track trends in the health of global oceans. The index employs 10 goals which represent a combination of social and ecological indicators, in keeping with the view that humans are an integral part of marine ecosystems. The index score for all of the world's oceans under national jurisdiction was 60 out of 100, 'indicating substantial room for improvement'. This assessment particularly identified marine species declines, habitat loss and fish stock depletions as being negative trends that significantly reduced many national index scores, as well as the global index score (Halpern et al., 2012). Whilst these concerns are voiced in a less compelling and alarming tone than those above, presumably as a means of more constructively engaging with policy-makers, they do represent a further scientific expression of concern about the state of our seas. It is worth noting, at this point, that the term 'ecosystem health' is frequently used but rarely defined, so it is defined for the purposes of this book as 'a measure of the structural and functional integrity, biological diversity and resilience of marine ecosystems coupled with their capacity to provide sustainable flows of ecosystem services'. As such, the concept of ecosystem health is strongly influenced by ecocentric values (Table 6, Chapter 3).

There are particular concerns about the health of and prognosis for tropical coral reefs, which are both exceptionally important and extremely vulnerable. Tropical coral reefs cover only around 0.15 per cent of the global marine area, but around 33 per cent of all marine fish species and 25 per cent of the total number of marine macro-species are only found on such reefs (Davidson, 1998, 5), and they provide around 10 per cent of global fish catches, on which many millions of people directly rely for their subsistence and livelihoods. They also provide exceptionally high levels of many other ecosystem services, such as shoreline protection, tourism income, potential pharmaceutical products and cultural values (Moberg and Folke, 1999). Given the high values attached to coral reefs, it is particularly worrying that more than 60 per cent of tropical coral reefs are recognised as being under immediate and direct threat from local pressures, such as fishing, coastal development and pollution, rising to 75 per cent when local pressures are combined with thermal stress, i.e. coral bleaching, related to climate change (WRI, 2011, 3). Coral reefs are therefore understandably a priority for MPA designation, but Mora et al. (2006) found that whilst 18.6 per cent of the global coral reef area was then covered by MPA designations, only 1.6 per cent of the global coral reef area was partially protected by MPAs that were effective in restricting some extractive activities, and less than 0.1 per cent was completely protected by MPAs which effectively prevented all extractive activities. Whilst there are particularly critical concerns about such tropical coral reefs, it is important to recognise that there are increasing societal and scientific concerns about the status and on-going declines of most, if not all, marine ecosystems, as the assessments above indicate.

There are also related concerns about the status of marine fish stocks. The percentage of assessed stocks around the world that are classified as overexploited rose to a high of 32 per cent in 2008, this being a 'cause for concern' (FAO, 2010, 8), though it declined to 30 per cent in 2010 (FAO, 2012, 53). Some have argued that the basis for such assessments leads to overestimates of the proportion of overexploited stocks and that the status of stocks is actually relatively stable (Branch et al., 2011, whilst others have argued that the FAO statistics actually mask some worrying trends and indicate 'that the world fisheries are on a dangerous course' (Pauly and Froese, 2012). This divergence is a recurring theme in the literature and is one that will be returned to in Chapter 3.

Increasing concerns about the state of marine ecosystems and fish stocks are not confined to official and scientific circles. In 2006, news channels all over the world reported predictions that all global fish stocks will have collapsed by 2048, based on Worm et al.'s (2006) extrapolation. The media coverage of this paper was extremely high profile, though this is but one example of growing media interest in the ecological effects of overfishing. A book identified by the *Washington Post* as one of the best, but also disturbing, that year was *The Unnatural History of the Sea* (Roberts, 2007), which eloquently and meticulously recounts the history of serial declines in fish stocks and the ecosystems of which they are a part. This is one of at least nine books released between 2002 and 2012 which focus on concerns about the ecological impacts of overfishing, along with other pressures. The feature-length documentary *The End of the Line* (2009) highlights similar concerns and also attracted a great deal of media attention. These are just a few of the more prominent examples of media coverage related to growing societal concerns about the ecological impacts of overfishing.

Disney also adopted the seas as a setting for popular films such as *Finding Nemo* and *Shark Tale*. These animations projected characters onto marine creatures, giving the impression that they were sentient and had families, relationships, aspirations and fears, just like humans, i.e. anthropomorphism. These films may have promoted the extension of identity and concern for animals out to sea, though perhaps not in the same way and to the same degree as *Bambi* once did in relation to terrestrial animals (Whitley, 2008). Nonetheless, the marine 'Bambi effect' is significant, being both a reflection of and a catalyst for growing societal concern about sea creatures and their marine homes, albeit concern that some would dismiss as sentimental. The release, by Warner Brothers, of the film *Happy Feet*, set in an anthropomorphised world of penguins, went one step further, including a plea to humans by its terpsichoreally talented star to end overfishing, this being a root cause of the penguins' plight. These films are certainly an improvement on the impression projected by *Jaws* (1975), in which a white shark was portrayed as a malicious and malevolent man-eater, perpetuating cultural perceptions of seas as cruel and foreboding, and unleashing decades of persecution on increasingly beleaguered shark populations. The feature-length documentary *Sharkwater* (2006), by contrast, portrayed sharks as magnificent creatures with crucial ecological roles, but also as hunted and endangered victims of human greed.

The 'noughties' (2000–2009), then, seemed to be the decade when society's concern about human impacts on our planet extended out to sea, particularly in relation to growing concerns about the wide-scale impacts of overfishing. The marine environment seems to have progressed from being the neglected Cinderella to being very much the *Zeitgeist*. This, however, may not yet be a cause for celebration, as the media love bad news stories, and the still growing concerns about declines in marine ecosystems continue, unfortunately, to provide plenty of these.

MPAs as part of the solution

Many of the contributions discussed above highlight no-take MPAs as a key means of addressing these concerns. No-take MPAs are marine areas where all fishing activities are permanently banned, as are all other activities that involve the removal of living and non-living resources, e.g. recreational angling, shellfish collection, sand extraction. No-take MPAs, also often referred to as marine reserves, are more strictly protected than partially protected MPAs, which provide for some activities such as recreational angling and fishing with static gears (traps, pots, set nets, etc.) and pelagic trawls (towed through the water column, not across the seabed). Many marine conservationists consider that no-take MPAs are the only 'real' or 'true' MPAs, as is discussed further in Chapter 2.

Worm et al. (2006) report findings which indicate that no-take MPAs can make an important contribution to the recovery of the biodiversity, productivity and stability of marine ecosystems impacted by overfishing. Roberts (2007; 2012) concludes his accounts of the increasing threats to marine ecosystems with a call for no-take MPAs to provide for 'the return of abundance' and 'the future of fish'. The *End of the Line* (2009) features interviews with Callum Roberts in which he highlights the need for no-take MPAs. This feature-length documentary was released in the same year as the eminent oceanographer Sylvia Earl was awarded a TED Prize for her wish 'that we all use all means at our disposal ... to ignite public support for a global network of MPAs, hope spots large enough to save and restore the ocean, the blue heart of the planet'. Most recently, the better enforcement of MPAs and achievement of related targets was highlighted as a means of improving the health index of oceans, as effective MPAs can address negative trends in the status of marine species, habitats and fisheries that will address multiple goals (Halpern et al., 2012).

It is important to remember that these contributions are underpinned by thousands of scientific papers on MPAs; over the period 1980–2012 over 5,000 such papers have been published and their numbers are rising at a still increasing rate (Figure 1), constituting another indicator of the growing interest in MPAs. There were, on average, less than 10 papers a year related to MPAs in 1980–1989, but this grew to an average of nearly 60 a year in 1990–1999 and over 300 a year in 2000–2009, rising to ~570 papers in 2012. These papers have highlighted the impacts of fishing on ecosystems and on particular habitats and groups of species, many also focusing on the potential of no-take MPAs to address such concerns.

These growing popular and scientific concerns about the declining state of marine ecosystems and the potential of MPAs to address them have been accompanied by scientific calls for a greater proportion of the world's seas to be protected. In 1998, 1,605 marine scientists signed a call for governments to protect 20 per cent of the world's seas from threats by 2020 (Holmes, 1997; MCBI, 1998). In 2001, 161 marine scientists signed a scientific consensus statement calling for a network of no-take MPAs to conserve fisheries and marine biodiversity (NCEAS, 2001). Most recently, 271 scientists called for the designation of a worldwide system of very large no-take MPAs as 'an essential and long overdue contribution to improving stewardship of the global oceanic environment' (GOL, 2011).

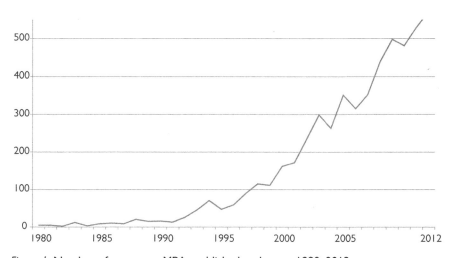

Figure 1 Number of papers on MPAs published each year, 1980–2012.

All databases on Thomson Reuters Web of Knowledge© were searched for all document types that include the term marine protected area, marine reserve, marine park or marine sanctuary (or plurals) in the title, keywords or abstract in each year.

International policy landscape

These calls are reflected by several international targets for MPAs, along with related targets for wider marine biodiversity conservation and sustainable fisheries management (Table 1). It is important to note, however, that such targets represent only a broad political commitment, in that only the Convention on Biological Diversity (CBD) is legally binding under international treatise law, the consequences of failure to meet even the CBD targets being more political than judicial. Nonetheless, such treatise do form an international framework for cooperation that creates political pressures to comply at a national level, and they are actually more influential than many people realise (Lyster, 1985, 14), though they do raise adjudicative and enforceability challenges (Sands, 2003, 11–15). There are also several other conventions under international treatise law that include provisions that can be applied to MPAs, such as the Convention concerning the Protection of the World Cultural and Natural Heritage (World Heritage Convention, 1972). These and other relevant conventions are reviewed in detail by Lyster (1985) and Sands (2003) and the World Heritage Convention will be considered later in this book in terms of its influence on relevant case studies. It is also worth noting that only the IUCN target specifically requires *no-take* MPAs, this being the most ambitious target in conservation terms but the 'softest' target in terms of its political significance. The other targets, including under the CBD, could be met by partially protected MPAs that include provisions for some extractive activities.

An analysis in the run-up to the Rio+20 conference (June 2012) highlighted that none of the targets listed in Table 1 have been fully achieved, though some progress towards some of them has been made (Veitch et al., 2012). This illustrates the relatively soft and non-binding nature of such targets under international treatise and agreements, other than through processes of political and public pressure, the publicity and accountability surrounding the Rio+20 conference over the failure to meet these targets being an illustration of the influential but non-binding nature of such processes. Overall, though, disappointment was expressed that the outcome of the Rio+20 document did little beyond reaffirming and reiterating commitments to existing targets, including the CBD target for 10 per cent of coastal and marine areas to be conserved through equitable and effective MPA designations by 2020. Recommitments were also made to the various fisheries-related targets though some deadlines will need to be expanded and efforts increased to achieve them (Table 1).

The United Nations Convention on the Law of the Sea (UNCLOS, 1982) is an important but neglected legal framework for MPAs. It provides the legal basis for delineating territorial waters, exclusive economic zones (EEZs) and continental shelves. UNCLOS also establishes the sovereign rights that coastal states hold over their waters, including the exclusive rights to exploit all living resources in EEZs. With such rights, however, come responsibilities, in that UNCLOS also obliges all states to 'protect and preserve the marine environment' (Article 192) and requires that coastal states 'shall ensure through proper conservation and management measures that the maintenance of the living resources in the EEZ is not endangered

Table 1 Main international targets for MPAs and fisheries

Platform	Target	Deadline
Convention on Biological Diversity (CBD), 10th Conference of the Parties (COP10, 2010)	At least 17 per cent of terrestrial and inland water, and 10 per cent of coastal and marine areas, especially areas of particular importance for biodiversity and ecosystem services, are conserved through effectively and equitably managed, ecologically representative and well-connected systems of protected areas and other effective area-based conservation measures, and integrated into the wider landscape and seascapes.	2020 – extends previous missed 2010 deadline
	All fish and invertebrate stocks and aquatic plants are managed and harvested sustainably, legally and applying ecosystem based approaches, so that overfishing is avoided, recovery plans and measures are in place for all depleted species, fisheries have no significant adverse impacts on threatened species and vulnerable ecosystems and the impacts of fisheries on stocks, species and ecosystems are within safe ecological limits.	2020
IUCN World Parks Congress, Durban (2003)	Establish a global system of effectively managed, representative networks of marine and coastal protected areas; these networks should be extensive and include strictly protected areas that amount to at least 20–30 per cent of each habitat.	2012
World Summit on Sustainable Development (WSSD, 2002) Plan of Implementation	Promote the conservation and management of the oceans through actions … including the ecosystem approach, the elimination of destructive fishing practices, the establishment of marine protected areas … including representative networks … and time/area closures for the protection of nursery grounds and periods.	2012 – MPAs deferred to 2020 CBD target
	Maintain or restore stocks to levels that can produce the maximum sustainable yield with the aim of achieving these goals for depleted stocks on an urgent basis.	2015
	Implement the FAO International Plan of Action for the management of fishing capacity to address overcapacity.	2005
	Implement the FAO International Plan of Action to prevent, deter and eliminate illegal, unreported and unregulated (IUU) fishing.	2004

by over-exploitation' (Article 61). The related UN Agreement for the Implementation of the Provisions of UNCLOS relating to the Conservation and Management of Straddling Fish Stocks and Highly Migratory Fish Stocks (UNSFA, 1995) provides for international cooperation to sustain such stocks under national and international jurisdictions.

Whilst UNCLOS does not explicitly and specifically require MPAs, it does provide the geopolitical and legal framework within which MPAs can be established. Strictly speaking, it could be argued that a coastal state whose EEZ includes deteriorating marine ecosystems and depleted fisheries, about which societal and scientific concerns are growing, is in breach of UNCLOS. The International Tribunal on the Law of the Sea is, however, more focused on settling boundary disputes related to sovereign rights to exploit living and non-living resources, leading to the neglect of obligations to conserve such resources under UNCLOS, these being deferred by default to other policy frameworks, such as the CBD and UNSFA. Nonetheless, UNCLOS has the potential to be an influential legal basis for obliging coastal states to restore and protect the living resources and the ecosystems that support them within their EEZs, though national political commitment to UNCLOS could be undermined were it attempted to actually realise such potential.

There are also several regional seas conventions and nine initiatives under UNEP's Regional Seas Programmes, started in 1972, that provide an important policy framework for conservation in some seas, some of which include MPA provisions. These are reviewed by Sands (2003, 399–458), to which the reader is referred for specific regional details.

MPAs in the context of sustainable development

The growing recognition of the need for MPAs must be considered in parallel with growing recognition of the need for sustainable development, as the two are inextricably intertwined. The UN Conference on the Human Environment (Stockholm, 1972) was significant in this respect, in that not only was it the first UN conference to recognise the need for MPAs, this being one influence that led to the IUCN's first international conference focused solely on MPAs in Tokyo, Japan in 1975, but also in that it laid the international foundations for what would eventually become known as 'sustainable development'. Recognition of the need to reconcile conservation and development had been growing for at least 100 years, in the light of growing awareness of the environmental consequences of the industrial revolution, coupled with a growing human population. The 1972 Stockholm Conference crystallised recognition that conservation and economic development were not mutually exclusive and that the aims of both had to be integrated if society were to become sustainable.

This recognition laid the foundations for the concept of what was to become known as sustainable development. However, this term did not become prominent until 1980, when it was employed as a central element of the IUCN *World Conservation Strategy*, which also recognised the importance and vulnerability of coastal and marine habitats. The UN subsequently established the World Commission on Environment and Development (WCED), which brought the concept of sustainable development to prominence with the publication of the Brundtland Report *Our Common Future* in 1987. This included the first widely recognised definition of this concept: 'development that meets the needs of the present without compromising the ability of future generations to meet their own needs.' Sustainable development was elevated to a central element of international policy debates as a result of the first UN Conference on Environment and Development (UNCED), also known as the World Summit on Sustainable Development (WSSD), held in Rio de Janeiro in 1992, where the UN Convention on Biological Diversity (CBD) was also formally agreed upon. A second WSSD was held in 2002 in Johannesburg. Both the WSSDs and CBD conferences of the parties (COPs) have established MPA and fisheries targets (Table 1). The status of marine ecosystems and fisheries and the role of MPAs remains a key priority and was a focus for discussions at the Rio+20 conference in 2012, though there were concerns that this would only lead to more 'empty ocean commitments' (Veitch et al., 2012) in terms of establishing more eventually unachieved targets (as in Table 1). As it turned out, Rio+20 merely re-affirmed commitment to these targets and came up with no new MPA or fisheries related commitments.

A brief history of MPAs

Marine areas have been set aside for millennia under traditional management systems, the Polynesians, for example, having long considered certain coral reef areas to be 'tabu', i.e. sacred and inviolate, though many other cultures around the world had similar traditions. Such set aside areas were probably based on a combination of spiritual, territorial and pragmatic considerations, including to restore fish populations and to supply surrounding reefs (Johannes, 1978). Whilst it is important to build MPAs on such customary practices (Cinner and Aswani, 2007), MPAs are currently recognised on a more formal, modern basis as 'Any area of intertidal or sub-tidal terrain, together with its overlying water and associated flora, fauna, historical and cultural features, which has been reserved by law or other effective means to protect part or all of the enclosed environment' (Kelleher and Kenchington, 1991, 11).[2]

The first such MPA was the Royal National Park, New South Wales, Australia, designated in 1879 (Anon., 2002), just seven years after the world's first 'modern' terrestrial protected area, Yellowstone National Park. The Royal National Park was predominantly terrestrial, though it included open coast and estuarine habitats. The first predominantly sub-tidal MPA was designated at Fort Jefferson, Florida, in 1935, covering the Dry Tortugas coral reefs but the first MPA based on first-hand appreciation of marine life, through the advent of snorkelling and SCUBA, is considered to be the Exuma Cays Land-and-Sea Park, designated in 1959 (Ray, 1999). Overall, the pace at which MPAs have been designated has been slow, compared to terrestrial protected areas (Figure 2).

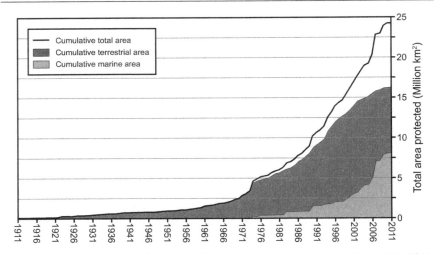

Figure 2 Cumulative area of terrestrial and marine protected area designations, 1911–2011.

Source: The World Database on Protected Areas, February, 2012 with permission of IUCN and UNEP World Conservation Monitoring Centre.

Slow progress and lack of effectiveness

The relatively slow progress of MPAs supports the view that 'the marine environment is the Cinderella of the conservation family, a family whose redoubtable energies for more than a century have been directed mainly to the land' (McIntyre, 1992). Whilst there has been much progress since 1992, when this statement was made, there is still much catching up to do. As of October 2012, there were around 10,000 MPAs, representing coverage of 2.3 per cent of the total area of global seas, including high seas, and 5.7 per cent of seas under national jurisdiction.[1] Whilst this represents a quadrupling of MPA coverage in the last 10 years, this still lags behind the 12.7 per cent coverage of terrestrial protected areas and falls far short of the CBD target for 10 per cent of the total marine area to be protected by 2010. This 10 per cent MPA target was subsequently extended to 2020 whilst that for terrestrial protected areas was increased to 17 per cent (Table 1). It is important to recognise that 60 per cent of this MPA coverage is represented by just 20 of the largest MPAs (TNC/UNEP-WCMC, 2012), which were recently designated and tend to be in relatively remote areas.

There is much variability between countries in terms of the number of MPAs, proportion of national marine area covered and type of MPAs (no-take through to partially protected), but one particular trend is that the number and coverage of MPAs is disproportionately high in more economically developed countries (MEDCs), and that less economically developed countries (LEDCs) are being 'left behind in the race to build a comprehensive global MPA network' (Marinesque et al., 2012). Given the growing concerns outlined above, the rate at which MPAs are being designated clearly needs to be increased. There is also a need, however, to ensure that MPAs are representatively distributed across all regions, including in LEDCs that host a very large proportion of endangered coral reef habitats. MPA coverage is highest in tropical realms (both MEDCs and LEDCs), at 6.15 per cent, compared to an average of 1.95 per cent and 2.31 per cent for northern and southern hemisphere temperate realms respectively (Toropova et al., 2010, 36). This relatively high MPA coverage is largely due to increasing recognition of the very high importance and vulnerability of tropical coral reefs and of the need for MPAs to better protect them, particularly where wider fisheries management approaches are lacking.

Designation is not enough, however, in that MPAs also have to be effectively 'protected' in order to achieve their conservation objectives. Gathering the information required to assess the number and coverage of MPAs is a major challenge, but it is one that is dwarfed by the challenge of assessing the effectiveness of many thousands of MPAs. An assessment of the management effectiveness of the then global total of 1,306 MPAs was undertaken in 1995 (Table 2). This found that there was only sufficient management information available to assess effectiveness for 29 per cent (383) of these MPAs, of which only 31 per cent (117) were generally meeting their management objectives, 40 per cent (155) partially meeting their objectives and 29 per cent (111) generally failing to meet their objectives (Kelleher et al., 1995). It is possible that a lack of management information is indicative of a weak management regime, in which case the total proportion of MPAs which have high management effectiveness will be less than 31 per cent. A worst case interpretation of the above figures is that only 9 per cent (117) of the total number of MPAs then around the world could be classified as having high management effectiveness.

Even if the MPAs for which sufficient management information was available are representative of the total number, the figures still indicate that 29 per cent of MPAs were failing to meet their management objectives, which clearly indicates that there is great scope for the improvement of MPA management effectiveness. An important caveat when considering Kelleher et al.'s (1995) review is that it did not address the level of protection which was afforded by the various types of MPA designation, other than an assessment of whether the management objectives were generally being met. Therefore MPAs which have modest management objectives might be classified as generally meeting these, whilst MPAs which have ambitious management objectives, and thus in need of more use restrictions, might be classified as not fully meeting these, despite the fact that the ambitious MPAs may be achieving more for marine nature conservation.

A guidebook of indicators for evaluating the effectiveness of MPA management has been produced (Pomeroy et al., 2004), but it has not been widely applied and there have been no global assessments of the effectiveness of MPAs since 1995, efforts being more focused on increasing their number and coverage, driven by targets (Table 1). The few evaluations that have been undertaken have been focused on MPAs covering tropical coral reefs, due to their particularly high importance and vulnerability, as discussed above. An evaluation of the effectiveness of coral reef MPAs in enforcing restriction looked at all (then) 980 such designations (Mora et al., 2006), covering 18.7 per cent of the global coral reef area. This found that less

Table 2 Management effectiveness of MPAs (after Kelleher et al., 1995)

Management level	Number/percentage of MPAs	
High: generally achieving management objectives	117	31%
Moderate: partially achieving management objectives	155	40%
Low: generally failing to meet management objectives	111	29%
Sub-total: MPAs for which sufficient management information was available	383	100% (representing 29% of total)
Unknown: insufficient management information available	923	71%
Total	1,306	100%

than 10 per cent of the area covered by these MPAs was effectively enforced, based on levels of illegal fishing (poaching), recognising that some fishing activities are legally allowed in most MPAs, and that only 0.5 per cent of the area of these MPAs was effectively enforced as no-take MPA. Another evaluation (Hargreaves-Allen et al., 2011) looked at the effectiveness of coral reef MPAs in improving conservation prospects, drawing on a sample of 78 such designations based on self-reports by managers. This found that 36 per cent of the MPAs showed no improvements in habitat quality, 28 per cent showed no improvements in species conservation and 44 per cent showed no improvements in fisheries management. Whilst these assessments of effectiveness may not be representative of MPAs covering habitats other than tropical coral reefs, they are indicative of wider concerns about the effectiveness of MPAs. There is widespread recognition that most MPAs suffer from a lack of effective management, and that a large proportion of MPAs are ineffective or only partially effective (Toropova et al., 2010, 7, 70).

This recognition of the lack of management effectiveness for many MPAs is the reason why this book focuses on governance. There are many other books that stress the need for improved marine conservation, including the crucial role of MPAs (e.g. Roberts, 2007; 2012a), and that provide excellent guidance on the ecological science behind the design and monitoring of MPA networks (e.g. Sobel and Dahlgren, 2004; Roff and Zacharias, 2011). This book complements these by focusing on the challenges of effectively and equitably governing MPAs. The improvement of governance incentives in a manner that steers human behaviour towards the achievement of strategic MPA objectives is the key to improving management effectiveness. This is considered in more depth in Chapters 5–8, but it suffices to say, for now, that the bottom-line for the assessment of the effectiveness of governance incentives in the MPA case studies will be the achievement of conservation objectives, i.e. the conservation and restoration of habitats, species and fisheries. The focus of the discussions, however, will be on how different incentives representing state, market and people-focused approaches to governance can be combined to promote the effective governance of MPAs.

Chapter 2

Objectives and categories of MPAs

Introduction

This chapter will introduce the main categories of objectives that MPAs are generally aimed at achieving, as this is a key driver of MPA governance. This is followed by a consideration of how the IUCN's Protected Area Management Categories can be applied to MPAs and of the increasing focus of many advocates on no-take MPAs.

The objectives of MPAs

The objectives of MPAs are extremely important as the differences between MPA objectives and the objectives of users of a given area are at the root of many MPA governance challenges. Whilst some local and incoming users may share the same objectives, or at least respect MPA objectives enough to cooperate with measures intended to support their achievement, other users may not. An often neglected issue with MPAs is that they are usually a vehicle for introducing or prioritising particular objectives in a given area, and many governance challenges arise from differences between the objectives of marine users and the objectives introduced or prioritised through the designation of a given marine area as 'protected'. This will be considered further in terms of value conflicts in the next chapter, but it is first worth introducing and reviewing the main categories of MPA objectives. The objectives of a particular MPA or network of MPAs will vary, depending on the context, the enabling policies and the lead agencies or interest groups, but broadly speaking MPA objectives tend to fall into nine inter-related categories.

Restore marine ecosystems

As was discussed in Chapter 1, growing scientific and societal concerns about the declining health of marine ecosystems since the mid-1990s have led to several calls for more of our seas to be set aside as no-take MPAs. Many papers have highlighted the impacts of fishing on ecosystems, e.g. Dayton et al. (1995); Pauly and Christensen (1995); Jennings and Kaiser (1998); Jackson et al. (2001); Thrush and Dayton (2002); Pauly et al. (2005); Worm et al. (2006); Reid et al. (2000). Some have particularly highlighted the structural impacts of fishing, from a top-down perspective, i.e. on higher trophic level apex predator and pelagic fish populations, e.g. Baum et al. (2003); Myers and Worm (2003), often discussed in terms of 'fishing down marine food webs' (Pauly et al., 1998) or 'trophic downgrading' (Estes et al., 2011). Other papers have highlighted the impacts from a bottom-up perspective, i.e. on lower trophic level benthic communities, particularly that heterogeneous, more structurally diverse and more productive seabed habitats tend to be reduced to homogeneous, less structurally diverse and less productive habitats, with a shift from more species diverse communities, including longer-lived, larger, attached fauna, to less species diverse communities, dominated by shorter-lived, smaller, mobile, scavenging, opportunist fauna (Watling and Norse, 1998). Such papers and related calls are motivated by concerns that the direct and indirect impacts of fishing are placing increasing pressures on marine ecosystems.

It is no coincidence that these papers and calls came at around the time of the emergence of the concepts of ecosystem-based management and ecologically coherent networks of no-take MPAs, as discussed later in this chapter, these being a response to such calls. The key focus of these concepts is that no-take MPAs are essential to restore ecosystems. This focus on restoring ecosystems remains important, so it is worth considering it in specific detail as arguably one of the central objectives of MPAs. There are three important concepts related to the MPA objective of restoring marine ecosystems: shifting baselines, trophic cascades and ecological resilience through species diversity.

Shifting baselines

Pauly (1995) has discussed how our perception of the state of marine ecosystems is based on our early baseline observations of seas already depleted by the impacts of fishing, and that this shift proceeds with every generation of scientists, masking the severity of progressive declines in the diversity of species and the abundance of populations: 'shifting baselines syndrome'. Dayton et al. (1998) have discussed this in terms of 'sliding baselines, ghosts and reduced expectations', the ghosts being now departed populations of species in a given area. More recently, Roberts (2007) vividly illustrated such downward sliding shifts in marine populations due to overfishing through studies of written and photographic records of the former abundance of fisheries now consigned to history books. Jackson et al. (2001) have called for initiatives to restore overfished populations and coastal marine ecosystems back to their natural state as revealed by palaeoecological, archaeological and historical records, rather than using such 'shifting baselines' as conservation targets. Overall, 'shifting baselines' has been a powerful concept for drawing attention to the long-term nature of declines in marine ecosystems and the urgent need for initiatives to restore them.

Trophic cascades

Trophic cascades are an 'ecological phenomenon triggered by the addition or removal of top predators and involving reciprocal changes in the relative populations of predator and prey through a food chain, which often results in dramatic changes in ecosystem structure and nutrient cycling'.[3] The early work of James Estes and his colleagues on the keystone role of sea otters in structuring kelp communities was influential in the application of this wider ecological concept to our seas. Their research revealed the trophic links through food chains between sea otters and kelp forests. Sea otters often break open and feed on the fleshy material of sea urchins. Declines in sea otter numbers through over-hunting, oil spills, etc. therefore lead to increases in sea urchins through reduced predation by sea otters. Sea urchins graze on kelp, so increases in sea urchin numbers suppresses kelp beds. This, in turn, suppresses populations of other fauna and flora that inhabit kelp forests but do not inhabit over-grazed sea urchin 'barrens', in a domino or 'trophic cascade' effect: fewer otters – more urchins – less kelp and related communities – reduced biodiversity (Estes and Palmisano, 1974). This was also discussed in terms of reduced expectations, as mentioned above, linking the trophic cascades and shifting baselines concepts.

Subsequent research revealed other factors were also likely to be perturbing kelp ecosystems, which function more as complex trophic webs rather than simple linear food chains, e.g. overfishing depletes fish populations – sea lion populations depleted due to shortages of fish on which they feed – killer whales normally often feed on sea lions, so killer whales switch to feeding on sea otters – more urchins, etc. (Estes et al., 1998). Depletions of populations and the local loss of other species which feed on sea urchins, such as spiny lobsters, cod and sheephead fish, have also been found to lead to similar trophic cascades and kelp ecosystem perturbations, undermining the diversity, stability and resilience of these important habitats (Steneck et al., 2002).

Trophic cascades are not confined to coastal kelp forests. Overfishing for cod and other large benthic zooplanktivorous fish has been identified as one of the key factors behind trophic cascades amongst phytoplankton, zooplankton, small pelagic zooplanktivorous fish and benthic macroinvertebrate populations in the open ocean eastern Scotian Shelf (Frank et al., 2005), as well as the semi-enclosed Black Sea (Daskalov et al., 2007) and Baltic Sea (Casini et al., 2008). The impacts of trophic cascades resulting from fishing on tropical coral reefs have also been highlighted (Roberts, 1995), as have those resulting from the loss of apex shark populations (Myers et al., 2007). It has also been found that trophic cascades induced by overfishing can perturb carbon flow dynamics, which could have profound consequences for the ecosystem functioning of temperate coastal ecosystems, particularly those in which primary production by macro-algae is important (Salomon et al., 2008), and that no-take MPAs can increase resilience to climate change induced phase shifts by restoring such carbon flow dynamics (Ling and Johnson, 2012). Other ecosystems are very likely to be affected by trophic cascades and related perturbations wherever significant depletions or the local loss of high trophic level populations occur as a result of fishing.

Given the historic nature of intense fishing, it is likely that many ecosystems may have started to undergo perturbations related to trophic cascades many centuries ago (Jackson et al., 2001). Trophic cascades initiated by fishing impacts may be one of the driving forces that push marine ecosystems into potentially irreversible regime shifts (Daskalov et al., 2007; Jiao, 2009). This could lead to shifted ecosystems in which, for example, higher trophic level finfish species become so depleted that their niche is permanently taken over by lower trophic level species, such as jellyfish. The risk of such trophic shifts was highlighted by Daniel Pauly in the late 1990s in his warnings of what could happen if we continue to 'fish down marine food webs' (Pauly et al., 1998), unleashing trophic cascades and regime shifts that could leave us little choice other than to eat 'jellyfish burgers', though at the time he considered this an absurd metaphor. However, in the light of growing evidence that jellyfish populations are increasing and that factors such as those discussed in this section could be contributing to such increases (Lynam et al., 2006; Brotz et al., 2012), the metaphor of 'jellyfish burgers' is now also becoming a salutary symbol, along with the more anglicised 'jellyfish and chips', of how we are changing our seas as a result of various impacts, including overfishing, and of the potential consequences of not providing for the restoration of marine ecosystems. Whilst there is evidence that recently observed increases in jellyfish blooms may actually be a consequence of long-term oscillations, rather than of degraded ocean ecosystems (Condon et al., 2013), such symbols have been employed in several campaigns for marine conservation measures, including networks of no-take MPAs.

More positively, *recovery* trophic cascades have also been observed in no-take MPAs, as populations of higher predators recover and more natural ecological structures and processes become re-established. One of the first studies of such recovery trophic cascades has been undertaken in the Leigh no-take MPA in New Zealand, established in 1979. The recovery of previously fished spiny lobsters and sparid fish that feed on sea urchins led to the reversion of over-grazed sea urchin barrens, dominated by crustose coralline algal habitats, back to macro-algal habitats, which eventually began to be colonised by other flora and fauna associated with these more productive and three-dimensional seaweed communities (Shears and Babcock, 2002). Some of the later indirect steps of the recovery trophic cascades were still being observed 25 years after their initiation (Shears and Babcock, 2003). Recovery trophic cascades resulting from no-take MPA designation have also been observed on tropical coral reefs (Mumby et al., 2006), the recovery of some large-bodied reef fish populations again being found to take decades (Russ et al., 2005; McClanahan et al., 2007). Micheli et al. (2004), Guidetti (2006) and Babcock et al. (2010) have undertaken meta-analyses of many examples of recovery trophic cascades related to no-take MPAs in temperate, Mediterranean and tropical regions. There is also evidence that recovery trophic cascades can eventually extend beyond the boundaries of no-take MPAs through the spillover of fish (Guidetti, 2007; Russ and Alcala, 2011). These and similar studies highlight the importance of complete and long-term protection for the indirect benefits of no-take MPAs, through recovery trophic cascades, to be achieved.

Ecological resilience through species diversity

Ecological resilience has been defined as 'a measure of the persistence of systems and of their ability to absorb change and disturbance and still maintain the same relationships between populations or state variables' (Holling, 1973). A common theme running through many of these studies is that marine communities with reduced diversity and depleted population sizes as a result of fishing and other impacts are less resilient to potentially disturbing or perturbing forces. The link between depletion trophic cascades, reduced species diversity and lowered resilience is a very important one. It is now widely accepted, after decades of debate amongst ecologists, that 'ecosystem stability is woven by complex webs' (Polis, 1998), in that higher species diversity, particularly the weak and strong trophic interactions between different functional groups of species (Figure 3), tends to lead to more stable ecosystems (McCann, 1998). Stability through the 'bracing' and reinforcement role of many diverse strong and weak trophic interactions tends to confer resilience to perturbing factors, such as ocean warming, species introductions and overfishing, which could potentially disrupt an ecosystem.

There are two inter-related ways in which diversity is recognised as conferring resilience:

- **Functional redundancy,** whereby different species within a functional group can adopt each other's ecological role should one species become depleted by overfishing, predation, disease, etc.
- **Response diversity,** whereby different species within a functional group display different responses to different environmental conditions, providing for adaptation to changing conditions (Bellwood et al., 2004; Elmqvist et al., 2003).

These mechanisms provide for the buffering of ecosystems against disturbance and for a diversity of adaptation responses. It follows that reductions in the species diversity and trophic complexity of marine ecosystems through the direct effects of overharvesting and the indirect effects of depletion trophic cascades driven by fishing will reduce both the diversity of functional groups and the diversity of species within functional groups, making these ecosystems less resilient to the perturbing effects of environmental change and other factors.

This has been demonstrated by many studies. Early studies on tropical coral reefs found that overfishing of apex predators and resultant trophic cascades can reduce the resilience of the ecosystem to shocks, such as hurricanes, and thereby potentially lead to regime shifts to alternative stable states with reduced diversity, dominated by macro-algae (Done, 1992; Knowlton, 1992; Hughes, 1994). These studies contributed to the growing recognition that human impacts on coral reefs are reducing

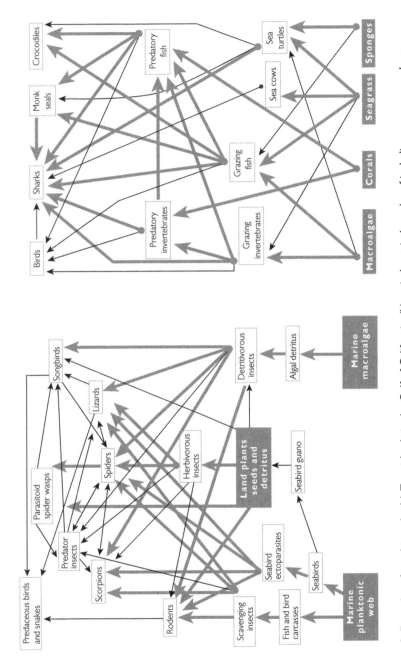

Figure 3 Examples of trophic webs (a) Two islands in the Gulf of California; (b) typical tropical coral reef including seagrass meadows (simplified). Nodes represent functional groups; thick arrows represent strong trophic interactions; thin arrows, weak interactions. (a) Adapted from Polis (1998) with permission from Macmillan Publishers Ltd: Nature; (b) Adapted from Jackson et al. (2001) with permission from AAAS.

both the diversity of functional groups and of species within functional groups, including population depletions, and thereby reducing the resilience of coral reef ecosystems, but that MPAs have the potential to maintain and restore resilience (Bellwood et al., 2004; Nyström et al., 2000).

Jackson et al. (2001) and Crowder and Norse (2008) highlight the particular importance of apex predators in marine ecosystems, along with structure-forming species such as coral reefs and kelp. They argue that human activities, particularly fishing, have been affecting marine ecosystems for many centuries, largely through the removal of apex predators and structure-forming species. Due to their wide ecological linkages, such species are 'strong interactors' in the trophic web, the abundance of which has strong direct and indirect effects on other species and on the structure and function of the ecosystem. The depletion or loss of such 'strong interactors' and, thereby, of interactions in the food web can have destabilising effects on trophic structure and function, undermining resilience in the ecosystem. Jackson et al. (2001) also found that there can be a time lag of decades to centuries between the onset of overfishing and the resultant ecological change, due to the buffering effects of functional redundancy, i.e., other species taking over the ecological functions of overfished species, until they too were depleted by over-fishing or other factors, leading to the eventual loss of entire functional groups and thereby to the reduction of resilience.

A meta-analysis by Micheli and Halpern (2005) of various ecological communities in the Channel Islands no-take MPA, along with data from 31 other no-take MPAs for temperate and tropical reefs, further investigated the factors that promote resilience in protected marine ecosystems. This found that functional groups of species impacted by fishing tended to consist of only one single species, depletions or the loss of which would reduce the functional diversity of the marine ecosystem as there are no alternative species that could take on the functional role of the depleted or lost species. Where more than one species constitutes a functional group, potentially providing for a degree of functional redundancy, they found that fishing tends to deplete or remove all the species within that functional group and thereby potentially removes that entire functional group. Therefore depletions of fished populations, either directly by being harvested or indirectly through trophic cascades, will reduce the species and functional group diversity of the marine ecosystem, and thereby reduce its resilience. It also logically follows and was found that no-take MPAs had a higher diversity of species and that this leads to a higher diversity of species within functional groups and of functional groups, which in turn leads to higher resilience (Micheli and Halpern, 2005).

These findings concur with an increasing number of other studies linking reductions in the diversity of marine species, including population depletions, through the effects of fishing, with reduced stability and resilience, and of the potential of no-take MPAs to restore resilience and the potential for recovery after disturbance, e.g. Hughes et al. (2003; 2005); Bascompte et al. (2005); Worm et al. (2006); Mumby et al. (2007); Sandin et al. (2008); Babcock et al. (2010); McCook et al. (2010); Mumby and Harborne (2010); Ling and Johnson (2012);

Bates et al. (2014); Howarth et al. (2014). Whilst many of these analyses are of tropical coral reefs, Micheli and Halpern (2005), Worm et al. (2006), Babcock et al. (2010), Ling and Johnson (2012), Bates et al. (2014) and Howarth et al. (2014) include temperate habitats. However, empirical evidence for similar resilience restoration effects for soft sediment habitats in tropical and temperate seas is lacking (Lester et al., 2009), though it is debatable whether this is due to the relative ease of showing such effects in reef habitats and a lack of no-take MPAs covering soft sediment habitats, or a lack of potential for such no-take MPA effects in soft sediment habitats, given their different ecological attributes (Caveen et al., 2012).

These debates aside, the inter-related concepts of shifting baselines, trophic cascades and rebuilding resilience underpin the MPA objective of restoring ecosystems. A key theme running through many papers is accordingly the potential of no-take MPAs to restore marine ecosystems impacted by fishing, e.g. Botsford et al. (1997), Roberts (1997); Watling and Norse (1998); Murray et al. (1999); Pauly et al. (1998; 2000), the papers cited above on the potential of MPAs to restore resilience providing further evidence for this potential. A global review of evidence of the biological effects of no-take MPAs confirmed that protection leads to significant increases in the biomass, density, species richness and size of vertebrates, invertebrates and algae within their boundaries, which represented recovery from the impacts of fishing (Lester et al., 2009) and which will increase the resilience of the protected areas. Such evidence further supports arguments that MPAs should be more about restoring marine ecosystems to rebuild resilience, rather than conserving them in a shifted baseline state, already impacted by the ecosystem impacts of fishing, including through trophic cascades. Such restoration is essential to rebuild resilience.

Restore and conserve marine fish stocks

Wider concerns about the impacts of fishing on marine ecosystems are accompanied by narrower but important concerns about the direct impacts of fishing on the stocks themselves. Fishing effort has rapidly increased in the last 200 years, through the advent of more and more powerful fishing vessels, as sail gave way to steam engines and then internal combustion engines, and the vessels became larger, deploying bigger nets and employing increasingly sophisticated technology to find fish. Fish landings have also increased, but the rate at which landings have increased has not kept pace with the rate at which effort has increased. Catch-per-unit-effort (CPUE) is a particularly important concept in fisheries management. In theory, this is a measure of the catch of fish for a given unit of fishing effort. Keeping track of global fish catches is a major challenge, as many catches are not recorded and reported.

It is a challenge, however, that is dwarfed by that of keeping track of fishing effort. Estimating catches is *relatively* straight-forward, in that you need only have to access the records or make estimates of only a few parameters, essentially the weight and first-sale value of landings for each species, recognising that some of the catch may have been discarded due to lack of quota, selection of higher priced sizes (topgrading), etc. Estimating effort, by contrast, involves measures of many metrics, the most commonly employed being engine power and time spent actively fishing, often expressed as kilowatt/hours (kWh). These alone are challenging enough to record/estimate, but effort is also affected by other factors, such as the size of the deployed gear (area of trawl mouth, length of longline/number of hooks, number of pots/traps, etc.), the design of the gear (which affects its efficiency in harvesting the target species), the electronic technology used to seek out the fish, etc. The on-going evolution of all these factors leads to 'technological creep', as the effort a vessel of a given size can exert slowly increases through innovation, at a rate which fisheries scientists estimate is proceeding at around 2–4 per cent per year, but reliably estimating effort for a given fishery that takes account of this and the many other factors that constitute effort is extremely challenging.

The one thing that is certain is that CPUE has decreased as the number and size of fish in a given stock is decreased, so more effort has to be expended to catch the fewer, smaller fish left in the depleted stock. A recent global analysis indicates that fishing effort has increased by a factor of 10 since 1950 and that the proportion of the global marine area that is being fished has risen from less than 10 per cent to nearly 70 per cent, but that fish catches relative to effort are half what they were in 1950, the downward trend in CPUE implying a downward trend in the abundance of global fish stocks (Watson et al., 2012). For example, it has been estimated that the CPUE for the seabed bottom trawling fleet in England and Wales in 2007 is 6 per cent of that in 1889, i.e.17 times more effort had to be exerted in 2007 to catch the same weight of fish as in 1889 (Thurstan et al., 2010).

The aim of conventional fisheries management is to avoid such stock declines and achieve the highest CPUE that can be sustained on a long-term basis, in order to provide for sustainable seafood supply and a profitable industry. This should be achieved by ensuring that fish stocks are maintained at a level which provides a maximum sustainable yield (MSY), recognising that the productivity of a fish stock can be increased by harvesting it but that an overfished stock cannot sustain an excessive harvest. This MSY level is roughly around 50 per cent of the unexploited 'virgin' biomass of the stock, though this varies very considerably depending on the attributes and dynamics of the stock in question. Maintaining stocks at a level that can maintain MSY involves regulating not only the amount of fishing effort that is exerted, but also imposing technical restrictions, such as minimum mesh sizes of nets and minimum landing sizes of fish, and restricting the total amount of fish that can be landed, in terms of quota. The concept of MSY has been referred to as the 'holy grail' of fisheries management (Larkin, 1977), as it represents a quest for something which some consider can never be found.

Even if it is feasible, there are many, many scientific factors that challenge, if not confound, assessments of what constitutes MSY. These confounding factors can be considered in terms of many huge questions (Table 3). All of these questions have to be considered in a sea of complexity, variability and uncertainty, as is discussed in Chapter 3. The challenging but fascinating science of addressing such questions

Table 3 Questions and related challenges that must be addressed in stock assessments

Questions	Related challenges
What is 'the stock'?	Identifying discrete fish stocks that do not significantly mix with other stocks
What is the virgin unexploited biomass of the stock?	Estimation is highly uncertain as stocks have generally been exploited for many years
What are the reproduction, growth and natural mortality dynamics, and thereby model parameters, of the stock?	Stock dynamics are complex and they vary amongst different species and amongst different stocks
What is the present biomass and age structure of the stock?	Challenges of counting, sizing and aging fish from a stock that may include millions of individuals that live under the waves and swim around
What proportion of the stock was harvested in previous years?	Challenges of accurately recording fish catches, bycatch in mixed fisheries, effort, etc.
What proportion of the stock can be harvested next year, based on last year's biomass and age structure?	Biomass and age structure will have changed in the meantime due to environmental conditions, ecological factors, fishing impacts, etc.
How does the stock interact with other stocks?	Estimating implications for stock dynamics where different fish stocks eat each other, compete for food, etc.
How does the stock interact with other non-target components of the ecosystem and how does this affect the dynamics of the stock and of the wider ecosystem?	Fishing a stock will directly and indirectly affect other ecosystem components and be affected by such components through processes such as trophic cascades

is considered in detail by many fisheries science publications, e.g. Jennings et al. (2001), to which the reader is referred for details.

An estimate of MSY must then be translated into the quota allocated to fishing vessels, which often raises political challenges, many regulators taking the upper figure from the range provided by the fisheries scientists, based on a range of confidence levels, and then adding to this figure due to pressure from the fishing industry and short-term political expediency. The quota then needs to be translated into catch limits for fishing vessels which are effectively adhered to, which often raises enforcement challenges related to bycatch in mixed fisheries, discards due to lack of quota, illegal and unrecorded landings, etc. The scientific, political and enforcement challenges of conventional fisheries management are massive and uncertainty is pervasive. The success of conventional fisheries management in addressing these challenges is variable amongst different stocks in different countries. Every two years the UN Food and Agriculture Organisation (FAO) compiles vast sets of data on various world fisheries trends, including the percentage of assessed stocks that are classified as overexploited. As was noted in Chapter 1, the proportion of stocks classified as overexploited has risen from 10 per cent in 1974 to a high of 32 per cent in 2008, this being a 'cause for concern', given that it was the highest such figure in the time series (FAO, 2010, 8), though overexploited stocks fell to 30 per cent in 2009 (FAO, 2012, 53). However, less than 20 per cent of global fish stocks are formally assessed and the FAO figures are derived solely from such data rich fisheries. Costello et al. (2012) assessed the status of the more than 80 per cent of fisheries that are not formally assessed, and estimated that 64 per cent of such fisheries were overexploited, i.e. below a biomass that could provide MSY. Even amongst assessed stocks, total annual world marine fisheries catches appear to have peaked at 86.4 million tonnes in 1996, subsequently declining to a plateau of around 80 million tonnes (FAO, 2012, 52).

Given that demand for fish is still increasing, that effort increases, including technological creep, should increase catches and that catches may only be being maintained by accessing new grounds and stocks (Watson et al., 2012), and by fishing down marine food webs (Pauly et al., 1998), this plateauing out and the potential for catches to decline, along with the high percentage of stocks that are classified as overexploited, has led to major concerns. Not least of all, the likelihood of achieving the WSSD target of restoring stocks to levels that can deliver MSY by 2015 (Table 1) is vanishingly small. Fish also provide 20 per cent of human protein requirements for 1.5 billion people, 400 million of whom critically depend on fish protein, particularly in LEDCs; coupled with a growing human population, the potential for fish catch declines also raises major food security concerns (Garcia and Rosenberg, 2010; Smith et al., 2010).

In the light of these trends and concerns, some have argued that conventional fisheries management approaches alone are evidently failing and that whilst efforts should continue to improve such approaches, they must also be complemented by networks of no-take MPAs, covering around 20 per cent of our seas (Holmes, 1997; MCBI, 1998; NCEAS, 2001). Used in combination with other fisheries conservation measures, it is argued that no-take MPAs can: (1) help restore stocks, enhance recruitment and sustain catches by the spillover of adults and the export of eggs and larvae into fished areas; (2) provide insurance and resilience in the face of uncertainty over MSY assessments, and enforcement problems, such as those outlined in Table 3; (3) protect habitats that are critical at certain stages of fish stock life cycles; and (4) provide areas for the study of fisheries and wider marine ecology (Botsford et al., 1997; Roberts, 1997b; Guénette et al., 1998; Lauck et al., 1998; Murray et al., 1999; NRC, 2001).

From the perspective of the MPA objective of restoring fish stocks, the spillover and export benefits of no-take MPAs are particularly important. The spillover of adults is expected to occur as finfish and shellfish grow in density (number of individuals in a given area or volume), size (length and weight of individuals) and biomass (total weight of the population), leading to adults swimming and crawling out of the no-take MPA during their natural movements or to seek less densely populated grounds. Once they have left the MPA, they will both boost the density, size and biomass of finfish and shellfish populations in the vicinity and be available for harvesting, boosting yields.

Increases in the export of eggs and larvae are expected to occur due to two factors. First, the growth in the density and size of finfish and shellfish within the no-take MPA should increase the number of sexually mature individuals, including a high number of mature 'broodstock' females. This is especially important given that for many fisheries, breeding may only be successful every few years when various conditions and factors coincide, so constantly having a much higher number of sexually mature individuals will increase the likelihood and productivity of intermittent breeding seasons (Longhurst, 2002). Second, big, old, fat, fecund female fish (BOFFFF) tend to have much higher numbers of eggs and larvae than smaller, younger females, as a result of the bigger size, health and maturity of the maternal finfish/shellfish (Marteinsdottir and Steinarsson, 1998). The eggs and larvae also tend to be larger, more robust and longer living (Morita et al., 1999), thereby having a higher survival rate (Berkeley et al., 2004). No-take MPAs allow many more such BOFFFF to build a broodstock, rather than females being caught as soon as they are big enough to be recruited to the fishery when they are often only just sexually mature, leading to a truncated population in terms of its age structure, which drastically reduces the numbers of BOFFFF.

The selective removal of larger fish by fishing can also exert an evolutionary pressure on fish populations, promoting the survival of slower growing, smaller individuals (Silliman, 1975). This can change the genetic structure of the stock, leading to a population with a genetic trait of more slower growing, smaller individuals, having negative consequences both for the health of the stock and its fisheries productivity (Law, 2000; Conover and Munch, 2002). Whilst no-take MPAs might provide for the maintenance of genetic variation that is unconstrained by such evolutionary fishing pressures (Conover and Munch, 2002), the main focus of their potential contribution to fisheries management is related to spillover and export benefits.

The rationale is that yields foregone through loss of access by fishers to the no-take MPA will eventually be more than compensated for by the increased yields of fisheries outside the closed area through spillover and export benefits, i.e. win-win for biodiversity *and* fisheries conservation objectives. Unfortunately, demonstrating such net win-win benefits through rigorous evaluations is extremely problematic, given that a sufficient number of comparable areas to provide for statistically adequate replication first have to be successfully closed to all fishing activities (Murray et al., 1999), often in the face of objections from fishermen. There are relatively few no-take MPAs and even fewer where before-after-control-impact (BACI) studies have been undertaken to objectively evaluate such benefits. Furthermore, the spatial heterogeneity and temporal dynamism of inshore marine ecosystems means that it is difficult to distinguish between MPA effects and wider habitat effects (Garcia-Charton and Perez-Ruzafa, 1999). The connectivity and variability of inshore marine ecosystems makes it difficult to confidently relate wider fisheries benefits to specific MPAs, particularly for migratory species.

Despite these challenges, the number of studies demonstrating the spillover and export benefits of MPAs is slowly increasing. There are many papers that report relatively predictable increases in the biomass, density, size and species richness of fish populations within no-take MPA boundaries, reviewed in meta-analyses by Mosquera et al. (2000) and Lester et al. (2009). There are fewer but a growing number of papers reporting increases beyond no-take MPA boundaries through adult spillover (Gell and Roberts, 2003; Molloy et al., 2009; Russ and Alcala, 2011) and the export of larvae (Harrison et al., 2012) and eggs (Crec'hriou et al., 2010), some papers including evidence of increases in CPUE around no-take MPAs as a result of such spillover and export, e.g. Roberts et al. (2001); Russ et al. (2004); Alcala et al. (2005); Murawski et al. (2005); Abesamis et al. (2006); Kellner et al. (2007); Goñi et al. (2008; 2010); Stobart et al. (2009); Vandeperre et al. (2011); Moland et al. (2013). It is important to note that these studies include both tropical and temperate fisheries, confounding claims that such 'benefits beyond boundaries' are confined to the former tropical coral reefs.

However, the number of papers reporting these benefits is still quite low, and whilst their number and rate of publication is slowly increasing, many fishing industry related interests remain skeptical of the potential for such benefits. The lack of reports demonstrating such benefits has led to calls for MPAs, especially no-take MPAs, to be established on a precautionary basis (Lauck et al., 1998; Murray et al., 1999). Fishing industry related interests tend to be even more skeptical of such calls, though many marine conservationists argue that MPAs must be part of the solution to the impacts of overfishing. This divide is discussed further in the next chapter but suffice it to say, at this juncture, that whilst the MPA objective of restoring fish stocks raises hopes along with both uncertainties and disagreements, it also remains both important and controversial.

Contribute to a MPA network

MPAs can be selected, designated and managed on an individual basis, but the focus has mainly been on selecting MPAs on the basis of network design principles, so that they can contribute to a wider network, as part of wider marine spatial planning approaches. Until recently, such networks were largely considered in terms of being representative, as an important means of conserving examples of all major habitat types within a biogeographic region, particularly rare habitats and habitats supporting rare species. Such representative MPA networks are typically based on a hierarchical habitat classification scheme, as a means of identifying and protecting the best examples of a range of habitats, each habitat being represented on a replicated basis by several MPAs. This not only provides for a safeguard against damage to an individual MPA leading to a loss of the best example of a given habitat type, but also provides for a wider network that spans different sub-regions. The rationale is also that protecting a representative range of habitats will protect the different species that each habitat supports, including rare species. MPAs are often selected in particular areas which are rich, in terms of the diversity of different habitats, as a means of maximising representation within a given MPA designation.

The first international IUCN conference on MPAs in 1975 called for the establishment of a system of MPAs that represented the world's marine ecosystems, and throughout the 1980s and 1990s the focus of IUCN resolutions and initiatives on MPAs remained on representative networks. The WSSD target in 2002 and IUCN target in 2003 were still both focused on representative networks (Table 1). One of the key themes of the 1995 international review of MPAs (Kelleher et al., 1995) was an assessment of the degree to which MPA designations represented the major biogeographic types in each of 18 marine regions, and the 2010 review of MPA progress also focused on representativeness, concluding that coverage in many ecoregions is still patchy (Toropova et al., 2010, 37), indicating that many habitats are under-represented and/or that coverage is too narrow.

In recent years, the focus on MPAs constituting representative networks has been supplemented by a complementary focus on MPAs constituting *ecologically coherent* networks. These are networks of MPAs within a biogeographic region that are 'connected by larval dispersal and juvenile or adult migration' (Lubchenco et al., 2003). This concept draws on research undertaken in the 1970s on the source-sink dynamics of larvae of the problematic crown-of-thorns starfish in the Great Barrier Reef Marine Park (GBRMP) (Kenchington cited in Dight, 1995). This research was undertaken as a means of informing the design of more strictly protected zones within the GBRMP. Given the extremely large size of the GBRMP, which covers ~346,000 km^2, such zones essentially constituted MPA networks. Building on such work, studies on larval transfer patterns were used to inform the design of coherent MPA network proposals in California (Carr and Raimondi, 1999).

The concept of ecologically coherent MPAs was first explicitly discussed in 1991, drawing on the terrestrial analogy of the dispersal of thistledown, stressing the need for networks of no-take MPAs that can influence recruitment in the wider marine area, including other no-take MPAs in the network, through the dispersal of larvae and eggs (Ballantine, 1991). It was later mentioned in 1994 in terms of MPA networks which can 'feed' each other with larvae (McManus, 1994). The concept of ecologically coherent MPA networks became more widely recognised in 1997 in a paper describing how connectivity patterns could be mapped, by assessing currents and thereby larval dispersal routes, as a basis for designing networks of Caribbean MPAs that provide for connectivity between sources and sinks of larvae of certain important species (Roberts, 1997a). In the same year, Ballantine (1997) highlighted the need for MPA network design to provide for some level of connectivity between MPAs, but stressed that such connectivity would be inherent in a network designed on the basis of biogeographical representativity and replication criteria.

There continues to be a divergence between those who consider the need for a detailed understanding of the dynamic patterns of currents, larval dispersal routes, fish migration routes, genetic population structure, etc., in a given region as a scientific basis for designing coherent MPA networks, e.g. Crowder et al. (2000);

Gaines et al., Shanks et al., Palumbi, cited in Lubchenco et al. (2003), and those that take a more pragmatic approach, on the basis that a detailed understanding of such patterns will rarely be feasible, instead employing representativity and replication criteria along with bet-hedging strategies to provide for coherence in the face of uncertainty, e.g. Ballantine (1997); Roberts (1998; 2000).

The former fear that designing MPA networks with only a partial or incorrect understanding of connectivity patterns could negatively perturb dynamics, through factors such as displacement of fishing effort from MPAs to areas that are sources of propagules. The latter argue that it is not feasible to understand anything other than a partial snap-shot of such dynamic patterns, which will change as a result of MPA networks, as well through natural variability (see Chapter 3), and that measures to designate MPA networks to address growing concerns about the diversity, health and resilience of marine ecosystems should not be delayed by the futile quest for a full scientific understanding of source-sink patterns in every biogeographical sub-region. This is an illustration of the two perspectives on MPA governance: top-down, expert science-based, and bottom-up, guided by expert and local knowledge (Jones, 2001), discussed further in Chapter 4.

There have been initiatives to design ecologically coherent MPA networks that draw on knowledge of the source-sink patterns of particular species, such as fish that are commercially important and other species that represent a conservation priority, e.g. GBRMP (Fernandes et al., 2005), Californian MPAs (Sala et al., 2002; Airamé et al., 2003), but the reality is that this information was often too limited to be a basis of network design, even in such well-funded initiatives where much research has been undertaken. Instead, representativity network design criteria were used, alongside available but limited scientific and local knowledge on source-sink patterns. 'Rules of thumb' were also used, such as the maximum distance between adjacent MPAs, but it is debatable, given increasing recognition of the limited range of populations of some species/sub-species (see Chapter 3), whether networks which comply with such arbitrary connectivity criteria should actually be termed 'ecologically coherent' (Jones and Carpenter, 2009).

Whilst debates as to whether MPA network design realistically can be based on connectivity patterns and dynamics continue, the reality is that the primary focus remains on representative networks, including replication across sub-regions and the inclusion of certain connectivity criteria, e.g. Rice and Houston (2011). In keeping with this pragmatic, integrated approach, the latest target for MPAs under the CBD is for 'ecologically representative and well connected systems' (Table 1) and guidance on designing MPA networks has been produced, which defines them as 'a collection of individual MPAs or reserves operating cooperatively and synergistically, at various spatial scales, and with a range of protection levels that are designed to meet objectives that a single reserve cannot achieve' (IUCN-WCPA, 2008, 12). Overall, it is widely recognised that MPAs should not be considered as individual entities, in that a key objective of any given MPA and of any initiative to designate MPAs in any given sub-region is to contribute to a representative and connected network, but not necessarily a coherent one.

Contribute to marine spatial planning

During my PhD research in the 1990s, I was particularly drawn to an essay presented at the First World Conference on National Parks in 1962. This was not only because this was the inaugural international conference on issues related to protected areas, that just happened to have been held in the year in which I was born, but also because its author, Professor G. Carleton Ray, revealed some very interesting foresights as to the many challenges that inshore marine conservation faced. As well as stressing the 'absolute necessity of setting aside unmolested study areas in the sea, "parks" in every sense of the word with all life protected', building on a similar plea he had made in 1956, he also proposed that spatial planning policies commonly implemented on the land be extended to inshore seas in order for different uses to be zoned, recognising the many pressures these different uses exert and the interdependence of land and sea (Ray, 1962, 78, 84–85).

It is logical that the association between the need for MPAs and the need to manage cumulative pressures on wider seas through the zoning of different uses persists. MPAs have long been seen by many as a means of developing and demonstrating the overall benefits of management approaches which enable multiple uses to co-exist on a sustainable basis in marine areas which are subject to a diversity of development pressures (Kenchington and Agardy, 1990). Multiple use, multiple objective MPAs were considered 'small scale models of the kind of integrated marine resource management which should be practised on regional and even global scales' (Agardy, 1994, 6). Management approaches which provide for multiple-uses through partial protection are particularly important in marine environments, as the provision of access and allowance of different activities is widely perceived as a reasonable expectation, compared to terrestrial areas where exclusion and strict protection is a relatively familiar and accepted conservation approach (Jones, 2001). The GBRMP has long been regarded as a flagship model of such spatial planning approaches that integrate conservation and different uses (Kelleher and Kenchington, 1982; Kenchington and Agardy, 1990; Kenchington, 1990), as it essentially represents spatial planning at a wider-scale, including no-take MPAs, the GBRMP being larger than many countries' entire continental shelf.

The related approach of marine spatial planning (MSP) emerged in the mid-2000s. MSP is 'a public process of analysing and allocating the spatial and temporal distribution of human activities in marine areas to achieve ecological, economic, and social objectives that have usually been specified through a political process' (Ehler and Douvere, 2007). It is often considered as being a means of implementing the concept of ecosystem-based management (EBM) (Douvere, 2008), which had previously emerged in the late 1990s, alongside the related concept of ecologically coherent MPA networks. MSP, EBM and MPA networks can be considered to have co-evolved as a response to growing concerns about the state of our seas and of the related need to implement the concept of sustainable development in the marine realm. EBM aims to maintain ecosystems in a healthy, productive and resilient condition, so that they can provide the services humans want and need, through an integrated approach to management that considers the cumulative impacts of different sectors at a wider scale, including land-based activities, on the entire ecosystem (McLeod et al., 2005). Many consider ecologically coherent networks of *no-take* MPAs to be an essential and central element of an ecosystem-based approach to the management of our seas, e.g. Botsford et al. (1997); Roberts (1997b); Pauly et al. (1998; 2002); Murray et al. (1999).

There is an interesting trend here, as MPAs were, until the late 1990s, considered as providing for multiple uses through zonation, including, where feasible, no-take zones. This is consistent with the view that MPA designations provide for both conservation and compatible multiple uses, including sustainable fish stock exploitation involving methods that are compatible with the conservation objectives for particular habitats and species. This is also consistent with the concept of sustainable development, with which MPAs co-evolved (Chapter 1), and with the related primary goal – protection and wise use – of MPAs agreed by the IUCN in 1988.[4] However, whilst MPAs used to be considered by many as areas for managing the cumulative impacts of multiple uses through zonation, MSP, including through ecosystem-based fisheries management (discussed in the next chapter), is increasingly recognised as being an extension of this role into wider seas. This trend, alongside growing concerns about the ecological impacts of fishing and the need for no-take MPAs to restore ecosystems (see previous section), has led to a shift in the focus from partially protected to no-take networks of MPAs. This is also a reflection of growing concerns that partially protected MPAs generally only have an extremely small proportion which are set aside as no-take zones, and that this undermines the achievement of many MPA conservation objectives (Brailovskaya, 1998; Prideaux et al., 1998).

Some consider this to be a retrograde step, as it decouples MPAs from the concept of sustainable development and undermines the potential for the acceptance of MPAs amongst many users of the seas. Others, however, consider this shift in emphasis from *partially protected* to *no-take* MPAs to be essential to address concerns about the ecological impacts of fishing. The emergence of wider-scale approaches to MSP could also be considered as supporting this shift to no-take MPAs, as the original role of MPAs in promoting and providing for multiple uses through zoning is now being fulfilled through MSP in wider seas, therefore MPAs should focus on constituting core no-take zones, recognising that no-take MPA networks are an essential and central element of ecosystem-based MSP.

Recent calls for very large ecosystem-scale no-take MPAs (GOL, 2011) arguably represent an issue in this respect. The aim of this campaign is that a network of 15 completely no-take MPAs that are each at least 150,000 km^2 (57,915 miles^2) is designated by 2022. At present, there are four very large no-take MPAs related to this campaign (Table 4) that alone represent 0.45 per cent of the total global marine area, or 1.17 per cent of the global marine area under national jurisdiction. Two of these MPAs, Chagos and Papahanāumamokuākea, along with the largest no-take MPA proposed around the Pitcairn Islands (834,000 km^2), are in overseas territories and it could be argued that this represents a return to a neo-colonialist 'fortress

Table 4 Very large no-take MPAs

Name	Location, Country	No-take Marine Area km^2	Designated
Coral Sea Marine National Park	Australia	502,654	2012
Motu Motiro Hiva Marine Park	Sala y Gomez Island, Chile	150,000	2010
Chagos MPA	Chagos Archipelago, British Indian Ocean Territory	640,000	2010
Papahanāumamokuākea Marine National Monument	Northwest Hawaiian Islands, United States	341,362	2006

conservation' approach (De Santo et al., 2011) that neglects the social justice implications of such very large MPA designations (De Santo, 2013a).

It could also be argued that the lack of adequate surveillance and enforcement capacity means that such massive designations essentially represent paper parks (De Santo, 2012), though the costs per unit area of designating and eventually actually enforcing such very large no-take MPAs could be less than those for smaller MPAs (McCrea-Strub et al., 2011) and early findings indicate that such very large no-take MPAs support a greater fish biomass across the community structure than smaller no-take MPAs (Graham and McClanahan, 2013). Despite such reduced costs and increased effectiveness, it is debatable whether designating vast marine areas as completely no-take MPAs is consistent with the concept of sustainable development. Very large but zoned MPAs that provide for small-scale fishing activities by indigenous human populations, as well as no-take zones, i.e., equivalent to wider scale MSP, could be considered to be more equitable without necessarily undermining effectiveness in achieving biodiversity and fisheries conservation objectives. The CBD target does not require that MPAs should be entirely no-take, but then the growing societal and scientific concerns about the declining state of marine ecosystems could be considered to be a sound basis for campaigns to encourage and support the designation of very large no-take MPAs.

When considering the objective of MPAs as contributing to MSP, it is important to recognise that MSP may not actually be 'ecosystem-based', therefore no-take MPAs may not be considered to be essential and central to MSP. Broadly speaking, there are two perspectives on MSP. Ecosystem-based MSP is underpinned by a 'hard sustainability' principle, whereby healthy ecosystems are considered the foundation of societal well-being, networks of no-take MPAs thereby being considered as an essential component of this foundation. Ecosystems are also often considered, in a related sense, to be fragile and vulnerable to human impacts, with associated risks of irreversible shifts or even collapses, MPAs potentially reducing these risks through promoting resilience.

Integrated use MSP, on the other hand, is underpinned by a 'soft sustainability' principle, whereby economic growth is considered the foundation of societal well-being, MPAs thereby being considered as a sectoral priority, amongst other priorities such as fisheries and tourism development, against which the need for MPAs can be traded-off as a priority (Figure 4). Ecosystems are also considered to be robust and resilient to human impacts, with a low risk of shifts or collapses and a high likelihood of recovery. There is also an associated faith that human ingenuity through technological development will overcome the consequences of environmental impacts and that economic development is necessary to provide for such technological development.

The reality is that most examples of MSP initiatives lie somewhere between these two extremes, but it is important to assess which perspective a given MSP initiative is tending towards when considering the feasibility of the objective of MPAs as contributing to such initiatives. Calls for ecosystem-based MSP within which ecologically coherent networks of no-take MPAs are essential, along with calls for very large no-take MPAs that cover marine areas that might be more suitable for wider-scale MSP, certainly represent a strong sustainability principle. Given the growing scientific and societal concerns about the declining health of marine ecosystems, it could be argued that a strong sustainability principle is urgently needed, rather than relinquishing marine conservation to integrated-use MSP, perhaps including a sparser network of partially protected MPAs with some small no-take zones, which could become a vehicle for continuing marine ecosystem declines through a business-as-usual approach. Such arguments are at the heart of the objective of MPAs as being an essential element, if not the foundation, of ecosystem-based MSP. Whilst it is widely agreed that an important objective of MPAs is to contribute to MSP, it is important to bear in mind that there are many different views on what MSP means and how MPAs should be designed to contribute to it.

Protect rare and vulnerable habitats and species

Whilst many papers have highlighted the impacts of fishing on particular groups of species such as sharks (Worm et al., 2013), sea turtles (Spotila et al., 2000), porpoises (Dalton, 2004) and seabirds (Verhulst et al., 2004; Croxall et al., 2012; Žydelis et al., 2013), the risk of marine species and the biogenic habitats they constitute becoming extinct was not, until recently, considered as being significant. Such optimism was largely on the basis that most species are widespread as a result of the wide scale and connectivity of marine ecosystems (see Chapter 3). Narrow endemism amongst marine species was therefore considered to be relatively low, making them less vulnerable to extinction than terrestrial species. It was also widely believed that species targeted for fisheries exploitation would become commercially non-viable long before they were at risk of becoming extinct. However, in view of the unprecedented degree and extent of pressures on inshore seas, and increasing awareness of the limited range of many species and of aspects of their life cycles that render them vulnerable to extinction, concerns about marine extinctions grew in the late 1990s.

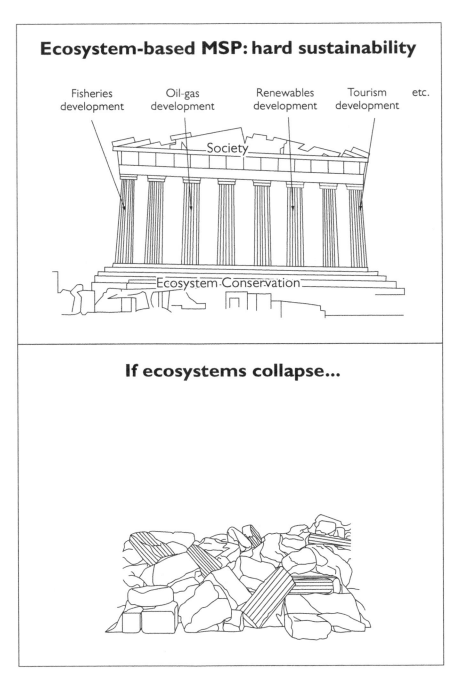

Figure 4 Hard and soft sustainability perspectives on marine spatial planning. Reprinted from Qiu and Jones (2013) with permission from Elsevier.

Integrated-use MSP: soft sustainability

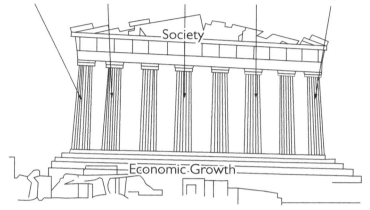

Fisheries development · Oil-gas development · Ecosystem conservation · Renewables development · Tourism development · etc.

Society

Economic Growth

If economic sectors and growth collapse...

These concerns were heightened when 133 cases of global, regional and local extinctions amongst marine populations were estimated to have occurred by Dulvy et al. (2003), due mainly to over-exploitation and habitat destruction, though the basis of some of these extinction estimates was questioned (Monte-Luna et al., 2007). In order to more systematically assess the risks of marine species extinctions, the IUCN initiated the Global Marine Species Assessment in 2006, as part of the global periodic assessment, the *Red List of Threatened Species*. Whilst this was on-going at the time of writing, 1500 marine species have been added to the *Red List* and assessments have been completed for six groups of marine species. These indicate that 17 per cent of shark and ray species, 13 per cent of groupers, 27 per cent of corals, 28 per cent of seabirds, 25 per cent of marine mammals and 86 per cent of marine turtle species are threatened (critically endangered, endangered or vulnerable to extinction) by a variety of impacts, particularly the direct (stock over-exploitation) and indirect (bycatch, habitat damage, trophic cascades, etc.) impacts of fishing, along with impacts from coastal development, pollution, ocean warming, etc. (Polidoro et al., 2008).

There is growing recognition and evidence that a potentially destructive wave of marine extinctions could be gathering force (Roberts and Hawkins, 1999) due to the growing extent, intensity and diversity of impacts on marine ecosystems. Where marine species and the habitats they constitute are being depleted and becoming threatened, MPAs, especially no-take MPAs, can help protect them against localised impacts, particularly related to fishing and coastal development, e.g. a small no-take MPA in South Africa was found to quickly benefit endangered African penguin populations by reducing their foraging effort, potentially increasing their breeding success (Pichegru et al., 2010). MPAs can increase the density and biomass of populations, which can also increase their resilience to wider-scale impacts related to ocean warming, pollution, etc., as was previously discussed in relation to the objective of ecosystem restoration. There can, however, be conflicts between this and other objectives. The restoration of marine ecosystems can lead to larger populations of predators that could deplete prey populations, including rare species. Restored shark populations may be one of the reasons for declining numbers of critically endangered monk seals in the Papahanāumamokuākea Marine National Monument (Gerber et al., 2011). On the whole, though, MPAs are recognised as positively contributing to the objective of protecting rare and vulnerable habitats and species.

Promote research and education

Scientific studies on marine ecosystems, including natural fisheries and ecosystem dynamics, need benchmark areas that are as unaffected as possible by the local impacts of human activities, such as fishing, recreation and waste disposal, as the effects of these could obscure the processes under investigation. Such benchmark areas, also referred to as 'pulse sites', can also be valuable for comparatively assessing the ecological impacts of harvesting in exploited areas and for monitoring wide-scale natural and anthropogenic changes in the absence of localised anthropogenic impacts (Murray et al., 1999; NRC, 2001, 27–28). No-take MPAs are an ideal designation for providing such benchmark areas.

Healthy, diverse marine ecosystems are also increasingly recognised as an important source of genetic resources, particularly of biochemicals that are of research value as potential pharmaceutical and biotechnology compounds. The need to recognise the objective of MPAs of protecting and providing such marine genetic resources has been identified as an urgent priority, as has the need to address related issues such as biopiracy and benefit sharing (Arrieta et al., 2010).

As well as such research values, MPAs also provide an ideal focus and vehicle for educational initiatives to raise public awareness and appreciation of the diversity and importance of marine life and of the threats to it. Many MPAs include interpretation centres that raise awareness not only of the importance of and threats to that given marine area, but also of growing concerns over the health of wider marine ecosystems and of the related need for improved marine conservation measures (NRC, 2001, 28).

Control the impacts of tourism and recreation

Tourism is one of the largest and fastest growing industries in the world, and coastal areas are a particularly important destination for millions of tourists, whether for relaxing on the beach or engaging in more active recreational pursuits, such as diving, snorkelling, angling and surfing. The growth of coastal tourism also places major pressures on inshore marine ecosystems, through the impacts of hotels and related infrastructure developments, increased demand for seafood, sewage pollution, damage by divers, etc. Ecotourism is often promoted as a solution, but this is akin to promoting sustainable development as a solution, as both concepts attempt to balance environmental, social and economic objectives, trade-offs amongst these objectives and their related impacts being inevitable.

Measures to manage the environmental impacts of tourism can range from the token, such as not changing towels on a daily basis, through to the more ambitious, such as eco-lodges and composting toilets. Environmental impacts are, however, generally only reduced and some degree of impacts from the growth of coastal tourism is inevitable, given that this generally leads to a very rapid growth in the human population. Even relatively 'eco' activities, such as whale watching and organised scientific expeditions involving paying volunteers, can still have significant environmental impacts. Coastal tourism is often focused on biodiversity hotspots, such as tropical coral reefs, which are particularly sensitive in a variety of ways, e.g. eutrophication from sewage leading to algae outcompeting corals, physical sensitivity of corals to touching, trampling and bumping by snorkellers and divers.

The development of coastal tourism in a given area tends to follow a general pattern, similar to that for the development of fish stocks, from discovery by incoming entrepreneurs to the rapid development of hotels, resorts, etc. and an influx of developers, often leading to the eventual degradation of the resource and a reduction in profits. This can also have significant social impacts, such as traditional communities and small-scale fishing livelihoods being displaced by the growth of tourism. Some local societal sectors may significantly benefit from such developments, particularly well-connected elites, but tourism industry workers tend to be trained and experienced incomers, limiting the potential benefits of tourism development to some marginalised local communities, who are often left with the low-paid menial jobs, if any. Large-scale tourism development is rarely driven by local communities, so many of the economic benefits tend to 'leak' away from locals to external developers, whilst the local economic benefits tend to be unevenly distributed. Ecotourism often includes measures to minimise the social impacts and maximise the economic benefits to all sectors of local communities, through employment, including training opportunities, as well as service and accommodation provision, but the effectiveness of such measures is often limited, even 'ecotourism' often being driven by incoming entrepreneurs, leading to the leakage of economic benefits.

Whilst there are growing concerns about such environmental and social impacts, tourism development can be vital to the growth of local, regional and national economies, often forming a very significant proportion of gross domestic product (GDP). Protected areas, including MPAs, can often be a focus for tourism development as the natural attributes of the area attract tourists, the designation of an area as 'protected' often being a vehicle for tourism development, by highlighting the natural beauty and wildlife interest of the areas. MPA status should thereby be a means of managing the impacts of tourism as well as promoting tourism, but the effectiveness of protective measures varies greatly.

There are growing concerns that protected areas are increasingly justified on such 'win-win' grounds, i.e., that tourism income exceeds the income from extractive activities, such as fishing, as well as being less environmentally damaging. Governments, corporate tourism developers and conservation NGOs are argued to sometimes work in collaboration to open up new areas to tourism development in 'protected' areas, often leading to the unjust exclusion of small-scale local users (Brockington et al., 2008, 146–147). This is also discussed in terms of an 'unholy alliance of global-level environmental and commercial interests' (Homewood et al., 2009, 25) and it is argued that even 'ecotourism' should be approached with caution in this respect (Brockington et al., 2008, 147). Such concerns also apply to MPAs, particularly the social justice impacts of MPAs promoted by conservation and tourism interests (Oracion et al., 2005; Fabinyi, 2008; Lucas and Kirit, 2009; Mascia and Claus, 2009), along with the environmental impacts of tourism development in MPAs. Despite such concerns, dive tourism-led entrepreneurial MPAs have recently been argued to have considerable potential, provided the related social and environmental impacts are carefully managed, including through the continuing regulatory and steering role of the state (Colwell, 1997; de Groot and Bush, 2010; Bottema and Bush, 2012; Fenner, 2012).

One particular environmental concern in relation to MPAs and the objective of controlling the impacts of tourism and recreation is that divers and snorkellers will damage the very attributes of the area that attracted them in the first place, thus endangering the sustainability of this economically important activity and undermining other uses (Kenchington, 1993; Davis and Tisdell, 1995). Management approaches such as diver education, zoning, carrying capacity and limits of acceptable change can be employed in MPAs to ameliorate such impacts (Davis and Tisdell, 1995; Roman et al., 2007), recognising that diver operators have a vested interest in employing such approaches to conserve the marine life that attracts their clients (Colwell, 1997). User fees can also be used to manage the numbers of divers, and they also have the potential to generate income to support MPAs (Depondt and Green, 2006). The impacts of other recreational activities can also be managed, e.g. catch-and-release angling (Cooke et al., 2006) and codes of conduct for whale watching, scientific research expeditions, etc., but concerns over the environmental impacts of 'non-consumptive' tourism and recreation activities in MPAs remain (Thurstan et al., 2012), including the potential impacts of recreational boat engine noise on the settlement of coral reef fish larvae (Holles et al., 2013).

MPAs clearly have a role in controlling the impacts of tourism and recreation, though they arguably also have a role in promoting such activities and being a vehicle for tourism development, potentially raising social justice and equity concerns. It is increasingly recognised that there is a need to systematically assess the effectiveness of 'ecotourism' and related management approaches in reducing environmental and social impacts and conflicts in MPAs (Dikou, 2010), in order to ensure that MPAs help achieve this objective, rather than undermining it. Given the continuing growth of tourism, particularly coastal tourism, the MPA objective of controlling the impacts of tourism and recreation is of particular importance, and will be considered further in the context of economic incentives (Chapter 7).

Maintain traditional uses

MPAs have long been considered a potential means of providing for the continuation of small-scale traditional uses, along with the exclusion of modern, market-economics driven exploitation, and thereby as a means of maintaining both natural and cultural heritage values, particularly in developing countries (Alder, 1996). MPAs that achieve this objective can also serve to maintain and demonstrate the value of customary approaches to managing marine resources (Johannes, 1978), provided MPA governance reinforces and builds on such approaches, rather than weakening and undermining them (Cinner and Aswani, 2007). To this end, the Convention on Biological Diversity (Article 10c) calls for measures to protect customary uses of biological resources that are sustainable. However, it has been argued that MPAs can subvert traditional resource management systems and promote market-economics driven exploitation by external interests (Nichols, 1999), thus undermining this objective. The discussions above, in relation to the objective of controlling tourism, on how MPAs may be a vehicle to promote tourism by incoming developers that displaces traditional uses, are another illustration of such concerns. Similarly, calls for large no-take MPAs discussed above, in relation to the objectives of promoting marine spatial planning and restoring ecosystems and fish stocks, could undermine traditional uses if no-take extends to banning small-scale subsistence fishing.

On the other hand, there are a number of papers which call the sustainability of small-scale 'traditional' approaches to fisheries management into question. Ray (1976) has argued that it would be a great error to categorise traditional subsistence cultures as right and the industrialised peoples as wrong, i.e. it should not be assumed that traditional cultures have an ecological basis. Polunin (1984) has similarly argued that traditional approaches to fisheries management based on tenure were competitively driven by a desire for gain rather than cooperatively driven by a desire for restraint. Cooke et al. (2000) report that the effectiveness of customary management in Fiji varied, some areas providing quite a high degree of management and others providing very little or no management. Jackson et al.'s (2001) findings reveal that overfishing by indigenous populations began to fundamentally alter some coastal marine ecosystems many thousands of years before the present, and they argue that these findings undermine the romantic notion of the supposedly superior ecological wisdom of non-Western and pre-colonial societies.

Even if it is accepted that some indigenous cultures are ecologically enlightened and can effectively manage fisheries, Ray (1976) argued that it must be recognised that traditional ways are subject to external influences and changing technology, particularly given the increasing influence of globalisation and related driving forces (Chapter 6). There are concerns, for example, that 'traditional' subsistence whaling actually employs modern equipment, such as exploding harpoons and fast powerboats (Yardley and Olsen, 2011) and that 'traditional' small-scale fishing increasingly employs large monofilament nets and powerful outboard engines, much of the catches being sold to national and overseas fish markets.

Such reservations about the sustainability of traditional approaches to managing marine resources contrast with the more recent focus on the importance of building on such approaches (Cinner and Aswani, 2007). Whilst such reservations must be recognised when incorporating the objective of maintaining traditional uses into the governance of MPAs, it is important that this objective is achieved where appropriate and compatible, though even traditional uses may need to be subject to a degree of restrictions in order to ensure effectiveness in achieving other objectives.

Cultural symbolic value of set-aside areas

This objective is derived from the 'moral conviction that it is right' to preserve natural areas and the species they support (Leopold cited in Pearsall, 1984), regardless of any scientific or resource management objectives. Indeed, Pearsall (1984) has argued that the likelihood of market values actually accruing from any given species is very small, therefore, public support for species protection must be founded on ethical perceptions, of which set-aside preservation areas are symbolic.

Set-aside areas have also been discussed as being the modern equivalent of pre-Christian sacred places in a society that has otherwise lost its links with the ecological community, though it is argued that there is a difference in the way that nature is regarded and the trade-offs that are permissible (UNEP, 1995). In a related sense, set-aside areas are also considered by some to be an important means of recognising the intrinsic value of nature and fulfilling the human race's stewardship duties on behalf of future generations and/or on behalf of God, i.e., dominion over the Earth comes with responsibilities. Those with neo-Marxist tendencies may wish to see areas set-aside from all capitalism-driven exploitation and as havens from globalisation.

Whilst such preservationist, ecocentric (see Chapter 3) and neo-Marxist motives all sound somewhat esoteric, such arguments can represent important moral motives for wanting to have marine areas set aside from all extractive activities, i.e. as no-take MPAs. Proponents of no-take MPAs will rarely, if ever, discuss such motives, but opponents of such designations have been known to accuse such proponents as being on a purely moral mission based on such motives (Jones, 2001). This divide is discussed further in the next chapter, but it is worth noting at this stage that symbolic moral objectives for setting marine areas aside as no-take MPAs can have a significant influence on related governance issues and debates.

Summary

This section has discussed the development of the thinking behind nine categories of objectives on which the designation and governance of MPAs can be focused:

- Restore marine ecosystems
- Restore marine fish stocks
- Contribute to an MPA network
- Contribute to marine spatial planning
- Protect rare and vulnerable habitats and species
- Promote research and education
- Control the impacts of tourism and recreation
- Maintain traditional uses
- Recognise the cultural symbolic value of set-aside areas.

From a governance perspective, it is important to remember that different people will focus on different objectives in relation to MPAs, and that there can, as was discussed at the start of this chapter, be conflicts between these objectives. More importantly, the introduction and prioritisation of MPA objectives, through the designation of a given marine area as 'protected', raises major governance challenges. The divergent views behind the conflicts related to these objectives are discussed further in the next chapter.

Different categories of MPAs

Along with the definition of MPAs and protected areas set out in Chapter 1,[2] the IUCN has also developed a scheme for categorising protected areas. Whilst this was primarily developed for terrestrial protected areas, it can be applied to MPAs (Table 5). Indicative guidance has also been provided on the activities, particularly different types of fishing, which might be appropriate in different categories of MPAs (Day et al., 2012). As most MPAs are zoned for different management priorities, different zones can be assigned to and reported under different management categories, provided the zones are formally delimited and their management priorities defined (Dudley, 2008, 56–57).

The categories that are most suitable for application to MPAs are 1a, II, IV and VI. There is increasing interest in applying category V to MPAs, due to its potential to provide for more flexible management in contexts where stricter protection is not feasible (Dudley, 2008, 21, 58). However, given the naturalness of marine ecosystems and that very few modifications resulting from human interventions, even 'traditional' small-scale sustainable fishing operations, are considered to lead to more varied and valuable assemblages of habitats and species (see Chapter 3), it could be argued that it is the flexibility of category V in allowing for existing activities that is attracting increasing interest, rather than the focus on preserving culturally modified coastal habitats.

It is important to recognise, as discussed in this chapter, that since the late 1990s there has been a shift in emphasis from partially protected or multiple use MPAs, consistent with categories IV, V and VI, to no-take MPAs, consistent with categories Ia and II. No-take MPAs, often referred to as 'marine reserves', are increasingly considered the only type of MPA that can achieve the objectives of restoring marine ecosystems and fish stocks, contributing to MPA networks and MSP, promoting research and education, and recognising the cultural symbolic value of set-aside areas. As was discussed in Chapter 1, all the scientific consensus statements calling for MPAs have specifically been for the no-take rather than partially protected categories, on the basis that only no-take MPAs can achieve these objectives, particularly the restoration of marine ecosystems and fish stocks that is necessary to address growing societal and scientific concerns about the state of our seas. These calls are supported by evidence based on meta-analyses of the effectiveness of MPAs (Lester and Halpern, 2008; Sciberras et al., 2014). These indicate that whilst partially protected MPAs may have some benefits over unprotected open access areas, no-take MPAs generally showed greater benefits, in terms of increases in the biomass, density, species richness and size of organisms within their boundaries, than partially protected MPAs.

However, most international targets for MPAs (Table 1) could be fulfilled by partially protected MPAs. The only target that specifies strict protection, i.e. requires no-take MPAs, is that established by the IUCN in 2003, but this is also the 'softest' target in that it has no legal mandate and is less politically influential than the other targets. Meanwhile, the reality is that whilst it is estimated that 5.7 per cent of the

Table 5 IUCN Protected Area Management Categories in relation to MPAs (shaded rows considered by author to be most applicable to MPAs)

IUCN Category	Application to MPAs (after Dudley, 2008, 55–58)	Particular primarily important objectives
Ia. Strict Nature Reserve	Focused on the preservation of biodiversity and other values in strictly no-take MPAs or no-take zones within MPAs	Restore ecosystems/fish stocks; contribute to MPA network/MSP; cultural symbolic value of set-aside areas
Ib. Wilderness Area	Focused on relatively undisturbed coastal wilderness areas, free of human disturbance and developments, with high seascape values; concept of marine 'wilderness' is less clear than for terrestrial protected areas	Control the impacts of tourism and recreation; cultural symbolic value of set-aside areas
II. National Park	Managed for ecosystem protection, providing for recreational and tourism but not fishing and other extractive activities	Restore ecosystems/fish stocks; contribute to MPA network/MSP; control the impacts of tourism and recreation; cultural symbolic value of set-aside areas
III. Natural Monument	Protection of particular natural features, such as seamounts, as well as cultural features such as submerged archaeological sites. Relatively uncommon marine designation	Promote research and education
IV. Habitat/Species Management	Protection of critical areas, such as breeding, spawning and feeding areas, that are particularly important for the conservation of priority species, and of habitats that need management interventions, such as preventing trawling and dredging	Protect rare and vulnerable habitats and species
V. Protected Landscape/ Seascape	Protection of culturally modified coastal areas that are characteristic of interactions of people and nature over time	Maintain traditional uses; cultural symbolic value of set-aside areas
VI. Protected Area with Sustainable Use of Natural Resources	Protection of natural habitats whilst allowing the ecologically sustainable collection of particular elements, such as particular food species or small amounts of coral or shells for the tourist trade[2]	Restore fish stocks; maintain traditional uses

marine area under national jurisdiction (excluding high seas) is covered by MPAs, only 1.4 per cent can be estimated to be covered by no-take MPA designations,[5] representing ~25 per cent of the total area of MPAs under national jurisdiction. Whilst this is a very large increase on the no-take MPA coverage of 0.2 per cent of the marine area under national jurisdiction by the end of 2006 (Wood et al., 2008), it still indicates that the majority (~75 per cent) of the area of the world's MPAs are designated for multiple use in that they are only partially protected. Furthermore, the vast majority of this no-take MPA coverage is due to recently designated very large no-take MPAs,[5] which tend to be in more remote areas, so most seas under national jurisdiction that are under pressure from nearby land areas with large human populations continue to have extremely low no-take MPA coverage.

Nonetheless, the increasing focus on and coverage of no-take MPAs arguably increases the potential for conflicts, underpinned by different ethical perspectives. It thereby also raises major governance challenges, as will be discussed in the next chapter.

Chapter 3

Differences and divergences

Introduction

This chapter will explore some key differences and divergences that raise challenges for the governance of MPAs. The first section will consider the different value priorities and related ethical perspectives, divergences between which are the basis of many MPA governance challenges. The second section will discuss an important divergence between marine ecologists and fisheries managers concerning different views on the role of MPAs and conventional fisheries management. The third section will review some key differences between marine and terrestrial ecosystems that raise some important MPA governance challenges.

Different value priorities

People perceive and value 'nature' in different ways (Table 6), the relative priority that is attached to these values varying and differing both amongst people and within a given person at a given time, depending on the context and the situation.

The importance of such underlying values is discussed by sociologists in terms of embedded normativities, recognising that these are very important in environmental discourses. The designation of a given marine area as 'protected', and the related prioritisation of a given set of MPA objectives, tends to inherently prioritise ecocentric and preservationist values over utilitarian values, and thereby raises basic conflicts (Jones, 2001). This is particularly the case for the objectives to restore marine ecosystems, restore marine fish stocks, protect rare and vulnerable habitats and species, and recognise the cultural symbolic value of set-aside areas, but there are also value conflicts related to the other objectives. The resolution of basic value conflicts that MPAs almost always raise may often not be easy or even feasible, as they are based on very different ethical perspectives, therefore 'negotiated settlement is foreclosed because consensus is philosophically intolerable' (Miller and Kirk, 1992).

One of the key challenges that MPA governance therefore generally needs to address, given that all MPAs entail a degree of protection, is such basic conflicts between utilitarian values, focused on sustainably utilising marine resources, and

Table 6 Different value priorities and ethical perspectives (after Callicot, 1991 and Jones, 1994)

Value	Ethical perspective	Basis	Focus of moral concern
Utilitarian	Resource Conservation Ethic	Based on Gifford Pinchot's arguments that nature consists of resources for human use that should be sustainably harvested	Material well-being of humans
Ecocentric	Evolutionary-Ecological Land Ethic	Based on Aldo Leopold's arguments on the importance of conserving the complexity and integrity of ecological processes and structures, of which humans are but an element	Well-being or 'health' of ecosystems
Preservationist	Romantic-Transcendental Preservationist Ethic	Based on John Muir's arguments that wilderness areas, set aside from exploitation, are necessary to fulfil our spiritual needs	Non-material well-being of humans

ecocentric and preservationist (henceforth collectively referred to as conservationist) values, focused on conserving ecosystem health and setting areas aside from direct human uses. This is, of course, assuming that all users of a given marine area recognise, at the very least, that marine resources should be sustainably utilised. Given the transient and mobile nature of many users of marine resources, the driving forces discussed in Chapter 5 that increasingly influence them, and the related potential for individualistic and short-term values and behaviour, even this assumption could be flawed, raising even greater basic conflicts and thereby even greater governance challenges. Chapter 4 also discusses issues related to the balance between imposing top-down MPA objectives on local users and allowing for them to emerge from the bottom-up. Accepting that MPAs must entail some degree of protection and therefore prioritisation of MPA objectives and related conservationist values, it is clear that basic value conflicts are at the root of many governance challenges.

Divergent views and the quest for common ground

Never the twain shall meet?

There are some very significant differences amongst marine scientists in their views on the status of marine fisheries and whether no-take MPAs are needed to restore fish stocks and the marine ecosystems that support them (Jones, 2007). The general pattern is that many fisheries scientists doubt the 'apocalyptic rhetoric' (Hilborn,

2010) of those who predict the collapse of marine fisheries and ecosystems, accusing them of faith-based advocacy of no-take MPAs (Hilborn, 2006), whilst many marine ecologists accuse such doubters of 'insouciance' (Pauly and Froese, 2012). This divergence of views amongst marine scientists over the gravity of the threats of overfishing has similarities to the 'alarmist or alarming' divergence amongst climatologists over the gravity of the threats posed by climate change (Risbey, 2008), the term 'denier' often being applied to those who emphasise the alarmist nature of warnings about such threats.

It is worth exploring the arguments of proponents and opponents of the need for no-take MPAs to restore fish stocks and ecosystems, as this divergence can have a significant influence on MPA governance processes, particularly on whether no-take MPAs are actually designated. Whilst recent papers that challenge the potential of no-take MPAs are still outnumbered by the many papers discussed in the previous two chapters that implicitly or explicitly make a case for no-take MPAs, the growing number of challenges is a significant trend that undermines the previous apparent 'consensus' amongst the scientific community on the importance of no-take MPAs (Jones, 2007).

It is increasingly being argued that conventional fisheries management approaches (CFMAs), such as quotas based on MSY estimates, effort reductions, partial seasonal closures and technical measures, as discussed in the previous chapter, will often be more effective than no-take MPAs for sustaining fish stock yields, not least of all because most fish stocks, apart from those associated with reefs, tend to migrate over wide geographic scales and are therefore inappropriate for protection through site-specific measures such as MPAs, e.g. Shipp (2003; 2004). Hilborn (2010) further argues that the apocalypse that many marine ecologists are warning of, e.g. Worm et al. (2006); Pauly (2009a), along with related calls for no-take MPAs, e.g. Pauly (2009b), has already been forestalled by improvements in CFMAs. This argument is supported by evidence that such improvements have led to the rebuilding of fish stocks (Worm et al., 2009), through improved CFMAs, including partially protected MPAs, closed only to certain fishing gears and/or over certain periods, and no-take MPAs. This paper was discussed as representing a 'détente in the fisheries wars' (Stokstad, 2009), as its first author was also the first author of the previous 'doomsday' paper predicting the collapse of all the world's fish stocks by 2048 (Worm et al., 2006) and it included many prominent fisheries scientists and marine ecologists amongst its other authors. It was also significant in that both CFMAs and MPAs, including no-take MPAs, were recognised as contributing to the rebuilding of global fisheries and the related recovery of marine ecosystems.

The foundations for this détente had been laid previously. Roberts et al. (2005) had recognised that CFMAs and no-take MPAs are complementary, and that one cannot be effective without the other as both have their limitations. Similarly, Hilborn et al. (2006) and Kaiser (2005) had previously argued that no-take MPAs and CFMAs will each be effective under certain conditions, and that the combined use of both approaches on an integrated basis should be explored.

These arguments appear to underpin and strengthen the détente, but there continues to be an important divergence, in that marine ecologists such as Roberts et al. (2005) are strongly advocating no-take MPAs on the basis that they represent a critically important way forward for restoring ecosystems, as do many others such as Murray et al. (1999) and Pauly et al. (2002; 2005). Whilst these authors accept that CFMAs will always have a role, they are strongly arguing that it is *essential* that a significant proportion of the world's seas is also set aside as no-take MPAs in order to provide for the restoration of marine ecosystems.

However, fisheries scientists, such as Hilborn et al. (2006), Kaiser (2005) and Shipp (2003; 2004) are primarily focused on the potential role of MPAs in sustaining fish stocks. Their assessments do consider the impacts of fishing on habitats and non-target species, but their primary focus is on improving the potential of CFMAs to promote sustainable fish stock yields. Many fisheries scientists also argue that the spillover and export benefits of no-take MPAs, discussed in the previous chapter, will be more than offset by fish yield reductions caused by loss of access to such MPAs. Overall effort and catches will therefore have to be reduced if significant areas of our seas are set aside as no-take MPAs (Hilborn et al., 2006; Shipp, 2003; 2004).

Along with arguments that natural refuges for fish already exist and that MPAs will displace fishing effort into such areas, damaging these previously unfished areas and increasing conflicts amongst fishers, many fisheries scientists continue to argue that the focus should be on the improvement of CFMAs, including the allocation of property rights to fish stocks (Chapter 5). Whilst some fisheries scientists accept that no-take and partially protected MPAs will continue to play a role, particularly for highly important and sensitive habitats, they are strongly arguing that the emphasis should be on improving CFMAs to sustain wide-ranging fish stocks, with no-take MPAs being pursued under certain conditions and on a cautious, experimental basis (Jones, 2007).

This focus on improving CFMAs is reflected by the shift towards ecosystem-based fisheries management (EBFM). EBFM represents

> a management system that is more holistic than single-species approaches by specifically considering indirect effects on habitats and food webs, protecting species that are incidentally captured and killed during fishing operations, acting to maintain valued ecosystem goods and services, and identifying potential conflicting management objectives and explicitly considering trade-offs among them.
>
> (Essington and Punt, 2011)

Whilst EBFM clearly represents a major improvement on single-species CFMAs, it is notable that the editorial to a recent special issue of a leading fisheries science journal on implementing EBFM (Essington and Punt, 2011) and a review therein (Rice, 2011) make no reference at all to the potential of MPAs as an element of EBFM. Similarly, a definitive paper on EBFM (Pikitch et al., 2004) recommends no-take MPAs as a precautionary measure only where there are low or moderate amounts of data, implying that such designations should not be necessary where data availability is good and EBFM can be fully implemented.

Norse (2010) considers that disputes amongst fisheries scientists and marine ecologists about the value of no-take MPAs in fisheries management 'seem destined to disappear' now that common ground has been established by Worm et al. (2009). This optimistic assessment is reinforced by other recent papers. For example, Gutiérrez et al.'s (2011) analysis includes MPAs as one of the key factors that promote successful fisheries management. Rice et al. (2012) report discussions at a recent international workshop amongst fisheries scientists and marine ecologists, one of the findings representing a convergence of views that 'spatial tools [such as MPAs] are essential for success in both fisheries management and conservation of biological diversity'. They conclude that 'MPAs can contribute significantly to increased reconciliation between fisheries management and biodiversity conservation'.

The détente (easing of relations) that has been established on the common ground amongst fisheries scientists and marine ecologists on the combined role of MPAs might be considered to be progressing towards an entente cordial (friendly agreement) and perhaps even towards a full alliance, whereby CFMAs and MPAs become integrated in terms of value principles, science, policy and practice. However, workshops such as that reviewed above, in which I was involved, often still involve the two sides 'talking past each other', in that whilst fisheries scientists and marine ecologists may think they are talking about the same thing, they are actually coming from a different starting point and want to end up at a different end point, any agreements that are reached tending to be misplaced or re-interpreted during the transit back to their disciplinary silos.

Different storylines and divergent views on addressing uncertainty

I have previously discussed this tendency to talk past each other in terms of parallel but contrasting storylines (Jones, 2007), an assessment of which reveals that whether one or the other perspective is adopted, the arguments can be considered logical, but the two perspectives will reach very different conclusions (Table 7). This is arguably the pattern that has emerged from the literature and related policy discussions on this broad issue, many publications arguing that no-take MPAs are a key way forward to complement CFMAs, as they are essential for providing for a precautionary and ecosystem-based approach, achieving wider ecosystem restoration gains, and addressing related societal concerns, and a growing number of publications arguing that CFMAs are the way forward, provided there is a sufficient knowledge base to inform it, with a few small no-take MPAs for research, education and very specific biodiversity conservation purposes. In the meantime, relevant politicians and regulators, as well as fishers and other interested parties, are receiving conflicting 'expert' opinions.

This, of course, is not unusual, as it applies to many other important environmental debates that are characterised by high scientific uncertainty, high

Table 7 Contrasting 'storylines' on the future of conventional fisheries management approaches (CFMAs) and the potential role of no-take MPAs (adapted from Jones, 2007)

CFMAs are failing due to	CFMA proponents	No-take MPA proponents
Poor enforcement	Co-management and rights-based approaches will increase fishermen's potential for cooperation, whilst increased technological and state capacity will improve enforcement.	No-take MPAs are easier to enforce, especially with vessel monitoring systems, as it simply has to be proven that vessels were fishing, not whether they were using a particular gear, catching a particular stock, etc.
Poor science	Increasing scientific expertise/technology, coupled with collective learning, adaptive management and more precautionary and ecosystem-based fisheries assessment approaches, will improve the scientific basis of decisions.	No-take MPAs have lower information requirements than CFMAs, i.e. less reliant on detailed knowledge of the status and dynamics of stocks as they employ simple design principles, e.g. habitat representativity and duplication in MPA networks.
Poor implementation of science	Regulators are not consistently and fully following scientific advice. Increasing cooperation amongst fishermen, scientists and regulators, coupled with increased political will to improve fish stock management, will increase the adherence to scientific advice on TACs, etc.	No-take MPAs are a simpler approach than CFMAs and their logical basis is clearer, so regulators and fishermen could potentially be more accepting of no-take MPAs and their scientific basis, provided they are willing to trial them.
Poor trust in science	Science will become better and this, coupled with collective learning amongst fishermen, regulators and fisheries scientists, will improve trust in the scientific basis of decisions.	The wider fisheries benefits that will *eventually* flow from no-take MPAs will *eventually* overcome mistrust of the scientific arguments for no-take MPAs. In the meantime, imposition will be required, provided sufficient political will exists, coupled with efforts to convince fishermen and regulators, and rigorous no-take MPA evaluations.
	Therefore CFMAs are the way forward, with a few small no-take MPAs for research, education and biodiversity conservation purposes ... but some fundamentally disagree with this perspective as they maintain a critical view of the positions adopted by CFMA proponents on these issues.	Therefore no-take MPAs are a way forward to complement CFMAs, as well as being essential for providing for a precautionary approach, and addressing wider societal concerns about declines in the health of marine ecosystems ... but some fundamentally disagree with this perspective as they maintain a critical view of the positions adopted by no-take MPA proponents on these issues.

potential environmental consequences and a high diversity of perspectives,[6] such as climate change (Risbey, 2008). It is important to recognise that scientific uncertainty is a key issue in marine ecosystems, as is discussed later in this chapter. Ludwig et al. (1993) have proposed that full scientific consensus concerning fisheries and marine ecosystems shall never be attained, as controlled and replicated experiments are impossible to perform in such large scale systems, and Jones (2006) reviews arguments that support this proposition in relation to no-take MPAs. Ludwig et al. (1993) have also argued that scientific uncertainty is not necessarily an obstacle to conservation initiatives, and that actions should be taken on an iterative, adaptive basis which recognises scientific uncertainty, rather than delaying actions in the quest for scientific certainty.

There are, however, divergent views between these two storylines on the role of science in addressing uncertainties (Jones, 2007) and of the way forward. Many proponents of CFMAs have argued that EBFM, including more precautionary approaches to setting TACs and recognising ecosystem interactions, can better address uncertainties. Many proponents of no-take MPAs have argued that such designations are necessary to address the uncertainties inherent in marine ecosystems and fisheries management, even if CFMAs move towards more precautionary approaches, especially given the need to restore ecosystems and build resilience.

Fisheries scientists tend to be more focused on maximising the sustainable yield of marine fisheries based on specific detailed studies, in order to feed a growing human population (Garcia and Rosenberg, 2010), whilst marine ecologists tend to be more focused on restoring marine ecosystems based on broader meta-analytical studies and 'rules of thumb', in order to address growing scientific and societal concerns about the state of our seas, though they do also recognise the importance of addressing food security challenges (Pauly et al., 2005). The latter's focus is often coupled with confidence that the spillover and export benefits of no-take MPAs will eventually lead to higher overall yields, this being a confidence or 'faith' that many fisheries scientists do not share. Given these different views on how to address uncertainties, translating Ludwig et al.'s (1993) arguments concerning the role of uncertainty in relation to the relative merits of no-take MPAs and CFMAs will raise major governance challenges.

Different value priorities amongst fisheries scientists and marine ecologists

By way of another illustration of the importance of embedded normativities, the divergences between these two storylines might also be a reflection of the different value priorities discussed above, no-take MPA proponents being more influenced by conservationist values, and CFMA proponents being more influenced by utilitarian values. These different value priorities and the related basic conflicts are the main reason why I am less optimistic than Norse (2010), in that it seems unlikely that the dispute between CFMA and no-take MPA proponents will disappear, as this dispute is underlain by different ethical perspectives, therefore 'consensus is philosophically intolerable' (Miller and Kirk, 1992). This underlying value divergence may also explain the sometimes rancorous tone of the arguments between fisheries scientists and marine ecologists.

In a related sense, marine ecologists are often advocates for marine conservation initiatives, such as no-take MPAs, in order to address growing societal concerns about the state of our seas, some fisheries scientists arguing that this has led to a loss of objectivity and a tendency towards 'apocalyptic rhetoric' (Hilborn, 2010). Fisheries scientists are often advocates for CFMAs, in order to address the concerns of fishers and fisheries regulators, some marine ecologists arguing that one of the reasons many fisheries scientists are 'insouciant' about the slide towards a marine dystopia may be that they have been unduly influenced by fishing industry interests (Pauly, 2009b; Pauly and Froese, 2012).

Given that marine ecologists and fisheries scientists tend to have different views on how to address uncertainties, different value priorities and different constituencies, it would seem that the prospects for establishing sufficient common ground to bridge these differences are less than good. Several recent papers do indicate that some common ground is being established (e.g. Worm et al., 2009; Norse, 2010; Rice et al., 2012), but there are also signs that it is being rapidly eroded.

This is illustrated, from the perspective of many fisheries scientists, by the papers discussed above arguing for a focus on improving CFMAs and, for instance, the complete lack of recognition of the potential contribution of MPAs to EBFM (Essington and Punt, 2011; Rice, 2011). Rice and Garcia (2011) go further in highlighting the 'conflicting pulls' between fisheries scientists, who they typify as being focused on addressing food security issues for a growing human population, and marine ecologists, who they typify as being focused on restoring marine ecosystems back to some pre-industrial idyll state through extensive no-take MPA networks based on a fortress conservation approach, with no serious analysis of the human costs. It is notable that Rice and Garcia (2011) make no reference to the potential contribution of no-take MPAs to addressing food security issues through spillover and export benefits. This paper also seems to represent something of a Malthusian analysis of despair,[7] in that it is pessimistic about the prospects for rebuilding stocks through CFMAs, even though rebuilt stocks could sustainably deliver a MSY that is higher than the present yield, as rebuilding stocks would impose costs through harvest reductions in the transition to former abundances that many human populations would not be in a position to bear, particularly those in poorer countries who depend on fish.

The critical view of fisheries scientists and CFMAs from the perspective of many marine ecologists is illustrated by the papers previously discussed that continue to argue that CFMAs alone cannot address growing concerns about the state of our seas, and thereby also continue to argue for no-take MPAs to restore marine ecosystems. Furthermore, the IUCN has recently published guidelines intended 'to make it more difficult for fisheries agencies to categorise area mechanisms that exploit fish as being MPAs'.[8] These contrasting perspectives between fisheries scientists and marine ecologists indicate that this common ground is being eroded by the eddying currents of these different views as fast as it is being established. Given that it is unlikely to be consolidated in the near future, scientific consensus on the need for extensive networks of no-take MPAs seems likely to remain elusive.

Finding a way forward

Despite this more pessimistic but arguably realistic analysis, it is important to constructively consider the prospects for moving forward in the face of continuing divergences, recognising that efforts to build as much common ground as feasible will continue. In this respect, it could be argued that the arguments of proponents of CFMAs are only valid only from a utilitarian resource conservation perspective, as they are only scientifically and ethically focused on sustaining fish stock yields. Proponents of no-take MPAs, on the other hand, could be considered to have extended their scientific and ethical concerns from fish stocks to wider fish populations, other marine species and the habitats and ecosystems that support them, reflecting growing wider societal concerns based on conservationist values.

Whilst EBFM aims to address such wider concerns by managing the ecosystem impacts of fishing, EBFM's primary focus remains on the utilisation of fish stocks, and proponents of no-take MPAs argue that only no-take MPAs can fully address wider societal concerns. As such, the arguments against the need for no-take MPAs put forward by proponents of CFMAs could be considered only to be valid in relation to the fisheries conservation objectives of such designations, i.e. such arguments do not extend to or undermine the non-target species, habitat and ecosystem restoration objectives of no-take MPAs. This represents an important caveat to the challenges to no-take MPAs such as those reviewed above, particularly given increasing recognition of the unpredictability of fish stocks and the ecosystems that support them, the fragility of marine ecosystems and the need for precautionary approaches in the face of uncertainty.

No-take MPAs have been promoted as a win-win approach, in that they can confer benefits for both marine biodiversity and fish stock conservation (Dalton, 2010). However, Ballantine (2002) has argued that the primary objective of no-take MPAs is to conserve or restore marine ecosystems and that whilst it is likely that some benefits will also be provided for fisheries, such benefits should be regarded as bonuses. Similarly, Halpern et al. (2004) have argued that no-take MPAs 'need not, and perhaps should not, be designed with fisheries management as a primary goal'. In keeping with these views, it is increasingly accepted that the primary objective of no-take MPAs is to restore marine ecosystems, with fish stock conservation and restoration objectives being secondary. The arguments against no-take MPAs based on their lack of fisheries benefits should accordingly be considered as only applying to the secondary fish stock conservation/restoration objectives of such designations and not to the primary marine ecosystem restoration objectives.

In a related sense, it has been questioned whether it is valid and necessary to 'sell' the fisheries management benefits of no-take MPAs (Jones, 2006), i.e. terrestrial conservationists do not have to convince farmers and hunters that protected areas will lead to increases in wider agricultural productivity and produce a surplus of wildlife that spills over and supports surrounding hunting communities, so why should we not think of no-take MPAs 'in the same way we think about terrestrial parks – simply as secure havens for biodiversity' (Kareiva, 2003). It is therefore arguably counter-productive for no-take MPAs to be 'sold' on a win-win basis, including their potential to deliver fisheries management benefits, as this detracts from their ecosystem restoration and related conservationist objectives, and leaves such calls open to arguments that CFMAs are better able to deliver fisheries management objectives.

Whilst such arguments against win-win approaches to selling no-take MPAs may seem valid in the context of a theoretical discussion, the advantages of selling the potential spillover and export benefits in overcoming objections from fishers are still compelling. However, fishers and their representatives are increasingly aware of the arguments against the potential for fisheries benefits based on spillover and export, such as those reviewed in this section. Furthermore, if such benefits do not eventually flow to a sufficient degree and extent to offset the foregone fish yields due to loss of access to a given no-take MPA, the win-win justification for the designation will be critically undermined, raising major governance challenges for the given no-take MPA and for other designations that are sold on a similar win-win basis.

Efforts to build common ground amongst fisheries scientists and marine ecologists are generally premised on the view that designating no-take MPAs and improving CFMAs should be considered as complementary approaches, in that both have important roles to play in addressing increasing concerns about the state of our seas. However, it is unlikely that a full alliance and consensus will be achieved given the divergence of views concerning the relative merits of no-take MPAs and CFMAs. The development of EBFM will hopefully also progress, though it is unlikely that this will proceed to the point that EBFM can fully address growing scientific and societal concerns about the declining state of marine ecosystems and the need to restore them.

Accepting that our seas are no longer regarded in purely utilitarian terms and that no-take MPAs are necessary to address wider societal concerns, it would seem to be more appropriate to pursue no-take MPAs on the basis that their primary objectives are focused on marine ecosystem restoration and related conservationist objectives. Many fishers understand that these are the most valid objectives for no-take MPAs, even if they do not agree with such designations or objectives. Fishers also consider that partially protected MPAs that allow compatible fishing activities, seasonal/partial closures specifically designated for fisheries management purposes, e.g. 'boxes' under the EC Common Fisheries Policy, and improvements in wider CFMAs are the most appropriate ways of improving fisheries sustainability (Jones, 2008). Where the restoration of marine ecosystems is concerned, however, it seems that explicitly focusing no-take MPAs primarily on this and related conservationist objectives is the way forward, fisheries restoration objectives being secondary or even coincidental objectives.

The case studies employed later in this book include some MPAs that are no-take and some that are only partially protected, as well as MPAs that have a variety of ecosystem restoration and related conservationist objectives, and fisheries conservation objectives, the balance between which varies. The related discussions concerning governance approaches in MPAs are premised on the view that MPAs represent designations that are necessary to address growing societal concerns about the state of our seas, recognising that some fisheries scientists and some people related to any given MPA may question this premise and that such questions will raise governance challenges. Whilst efforts to build common ground between marine ecologists and fisheries scientists will no doubt continue to make progress, the divergences between these two perspectives are also likely to continue to pose broader governance challenges related to MPA debates and initiatives.

Some key differences between marine and terrestrial ecosystems

Prior to the 1970s, the vast majority of protected areas were terrestrial designations (Figure 2). When extending what was predominantly a terrestrial designation out to sea, it is important to recognise that there are a number of ecological and management differences between marine and terrestrial environments which raise some significant governance challenges.

Three-dimensional

The seas are a three-dimensional environment, in that life can inhabit different depths from the surface to the seabed. Of course, terrestrial ecosystems do extend into the third dimension, e.g. trees grow towards the sky, pollen drifts in the wind and birds and insects fly, but the third dimension is much more significant in marine ecosystems. This is particularly the case given the high density of seawater, compared to air, and thereby the capacity for marine organisms to float and remain suspended at a particular depth with no significant effort. The average depth of the world's oceans is around 3.682km over an area of 362 million km^2, giving a total ocean volume of 1.3324 billion km^3 (Charette and Smith, 2010) The entire ocean volume represents a biome or potential space for life, which is why it is estimated that around 95 per cent of the planet's living space is beneath the oceans' waves. This book is focused on marine areas under national jurisdiction, which tend to be on continental shelves where the depth is generally no more than 150m, increasing if the area under national jurisdiction extends out to the continental slope, but this still represents a much more three-dimensional environment than on land. This is significant because protected areas were initially predominantly developed as a terrestrial designation, with the emphasis being on two-dimensional approaches to management measures, such as zoning. When extending such designations out to sea, the third dimension needs to be considered, Ray (1962, 81) long ago having recognised the necessity and challenges of designating three-dimensional MPAs.

As was mentioned in the previous chapter, the IUCN's protected area classification scheme can be used to classify zones within a given MPA, as well as to classify the entire MPA (Table 5). However, whilst there have been a few MPAs that have been formally vertically zoned, e.g. to allow pelagic fishing in the water column but to prohibit demersal trawling on the seabed, applying different zone classifications to different depth layers within a MPA is strongly discouraged by the IUCN. This is mainly because there may be ecological linkages between benthic and pelagic systems, such as the trophic cascades discussed in the previous chapter, so the exploitation of mid-water or surface fisheries could have ecological impacts on seabed communities. Vertical zoning is also considered to lead to enforcement challenges, as vessels fishing in the water column would have to be distinguished from those fishing on the seabed, as well as to challenges in entering such vertically zoned MPAs into existing databases (Dudley, 2008, 56). That is not to say, however, that different management prescriptions cannot be applied to different depths, e.g. to allow certain pelagic fisheries but to ban all seabed fisheries, but there is a strong presumption that these different management prescriptions should not used as a basis for separately categorised vertical zones (Day et al., 2012, 26). Whilst the logic behind these rationales is convincing, some have argued that the presumption against vertical zoning is an inappropriate extension of two-dimensional thinking from the terrestrial into the marine realm.

This aside, the three-dimensional nature of our seas presents governance challenges as it is no longer appropriate to think of MPAs as two-dimensional designations, in that features of conservation interest, potential impacts on them and measures to manage such impacts need to be considered at different depths.

Scale-connectivity

One of the consequences of the density of seawater and the buoyancy it thereby confers is that marine organisms can potentially drift and/or swim over very long distances during their larval, juvenile and adult stages. It was thereby assumed that most marine species and their constituent sub-species and populations had a very wide range, rather than being confined, i.e. endemic, to a particular biogeographical locality. Protected areas are, by their very nature, locality-specific designations and it has been argued that their effectiveness is limited in relation to the marine environment compared to the terrestrial environment, as MPAs cannot protect species, sub-species and populations whose wide natural range takes them beyond MPA borders. This is consistent with the arguments discussed in the previous section of this chapter that most fish stocks migrate over too wide a scale for MPAs to be effective in conserving them. Furthermore, if habitats and species in a particular locality were to be damaged, the scale and connectivity of marine ecosystems means that such damage would not be significant given the wide distribution of habitats and species, therefore site-specific designations such as

MPAs are not appropriate, as was discussed in the previous chapter in relation to rare species.

There are, however, growing challenges to such arguments about the spatial limitations of MPAs, particularly in relation to the importance of certain critical marine habitats as, for example, feeding and nursery grounds, that are required to support certain stages of the life cycles of otherwise wide ranging species. More recently, there also has been increasing recognition, through genetic studies, that the species of many marine functional groups have dispersal distances comparable to, or considerably less than, many species in comparable terrestrial functional groups (Kinlan and Gaines, 2003). There is also increasing recognition that many marine species, particularly commercially exploited fish that were assumed to have very wide dispersal distances, have sub-populations that are genetically adapted to local conditions (Hauser and Carvalho, 2008). Such findings are highlighting the importance of the need to protect species, particularly their sub-species and populations, and habitats in particular marine localities through MPAs, including inter-connected networks, in order to conserve the diversity and resilience of marine ecosystems.

Similarly, it has been assumed that MPAs cannot protect marine life within their boundaries from 'up-stream' pollution impacts and the impacts of wide-scale environmental changes, e.g. climate change. These assumptions are increasingly being challenged on the grounds that marine areas that have been destabilised by local impacts are more vulnerable to wider-scale impacts such as those related to climate change (Walther et al., 2002), including increasing evidence that MPAs can increase resilience, such as the many papers concerning 'resilience through diversity' discussed in the previous chapter. There is also an increasing focus on scaling-up MPAs by designing ecologically coherent networks that accommodate the scale-connectivity of marine ecosystems, as was also discussed in the previous chapter. Whilst such challenges and developments support the role of MPAs, it must be recognised that the scale-connectivity of marine ecosystems will present challenges to MPA governance, as people may question the potential effectiveness of such designations given arguments and assumptions about the spatial limitations of MPAs.

Variability

Marine ecosystems tend to be more complex than terrestrial ecosystems, due to their wider diversity of niches and greater number of trophic levels, marine ecosystems tending to have five trophic levels whilst terrestrial ecosystems tend to have three. Different communities with non-linear population dynamics tend to interact over larger spatial scales, due to scale-connectivity, and are closely coupled to physical oceanographic variations that occur on a cyclical or complex basis over scales of days to decades (Steele, 1991). Biological communities in inshore seas therefore tend to exhibit particular variability or discontinuities due to a combination of biotic, abiotic and anthropogenic factors, the interactions between which are increased by the connectivity discussed above (Ray, 1996). These factors lead to relatively variable and unpredictable marine ecosystem dynamics, compared to terrestrial ecosystems.

Uncertainty

The challenges of variability are exacerbated by uncertainty. Our understanding of the structure and function of marine ecosystems is poor compared to that of terrestrial systems, due to logistical problems of observing and studying such environments, the related high costs and the fact that humans are a predominantly terrestrial species. The combination of complexity, variability and poor understanding poses major challenges for MPA governance, due to difficulties in providing an evidence-base for proposed user restrictions. Such restrictions generally need to be scientifically justified through the determination of the significance of observed changes and identification of cause–effect linkages between the impacts of certain human activities and observed changes. Ideally, governance decisions should be underpinned by shared confidence in knowledge concerning the significance of such effects and the human activities that are their causes.

This is, however, rarely the case and both the significance of observed effects and their linkages to certain human activities often remain uncertain, given the complexity and variability, and difficulties of studying marine ecosystems, even if a great deal of funding is invested in research. For example, Dalton (2005) highlights how the potential link between overfishing and Steller sea lion declines is still being contested despite over $120 million worth of research, and Pratchett (2005) discusses how the causes of outbreaks of crown-of-thorns starfish in the Great Barrier Reef Marine Park are still not known despite much research. Even where such causal links between observed effects and particular human activities are claimed to have been established, they are often contested, particularly by those users whose activities might be restricted in order to address these effects. There will often be other scientific evidence from other experts that challenges claimed causal links and these different expert views will often be seized upon by user groups to challenge any restrictions that could affect them. Similarly, the export and spillover benefits of MPAs in replenishing surrounding fisheries can be contested on the basis that they are due to other factors and to variability, as was discussed in the previous chapter. The divergences between fisheries scientists and marine ecologists can be significant in this respect.

The tendency that the evidence base for restrictions is often uncertain and can be contested is the cause of many MPA governance challenges. Ludwig et al. (1993) have argued that scientific uncertainty is not necessarily an obstacle to conservation initiatives and that action should not be delayed in the quest for scientific certainty, as discussed above. However, translating such arguments into reality in the face of challenges to the evidence base for MPAs from users affected by restrictions is often very challenging, uncertainty often being the basis for many delays and compromises in designating and governing MPAs.

Hidden and alien

To the majority of people marine ecosystems are 'out of sight, out of mind' in that the impacts of human activities on marine habitats and species are hidden beneath the waves. For instance, a person observing a demersal fishing vessel from a cliff top would be unaware of the impacts of trawling on the seabed and the species it supports. By contrast, a person observing the effects of a tractor-towed plough on a grassland habitat or of tree felling can directly witness the impacts of human exploitation on terrestrial habitats and species and is more likely to be concerned. Whilst charismatic marine megafauna such as whales and dolphins do attract public attention, our direct observations of such species and appreciation of human impacts on them is relatively limited. These observational difficulties also mean that we do not become familiar with marine 'landscapes' and therefore cannot appreciate the changes that occur to them as a result of wide-scale and long-term human impacts in the same way that we become familiar with and come to care for terrestrial landscapes. This lack of historical understanding and appreciation of past marine landscapes and their associated communities of fish and other populations contributes to the 'shifting baseline syndrome' (Pauly, 1995), whereby we come to accept the current depleted state of fish stocks and wider biodiversity as we have few historical and cultural records of the thriving and bountiful nature of our seas in the past (Roberts, 2007).

Even if people are made aware of human impacts on marine ecosystems, their reaction could be one of indifference given the widespread view that the seas are an unfamiliar, alien world dominated by cold blooded animals that routinely abandon their young and amongst whom cannibalism is common. Marine populations tend not to follow familiar seasonal patterns and the sea itself is also often seen as an adversary – 'the cruel sea'. Certainly, humans do not have the 'hard wired' draw towards marine landscapes that some evolutionary biologists think we may have developed towards certain terrestrial landscapes (Ulrich, 1993), arguably due to a lack of familiarity and associations with marine landscapes and a lack of empathy with most marine life. For such reasons it could be argued that we do not appreciate our seas as consisting of living landscapes, areas of which might be preserved for our aesthetic and symbolic appreciation. Whilst some people particularly value marine life due to its unpredictable, mysterious and unusual nature, the majority of people are relatively unfamiliar with marine life and landscapes. Therefore gaining understanding of the need for use restrictions to protect marine life and gaining support for related restrictions presents significant challenges in MPA governance.

Naturalness

There are important differences between terrestrial and marine ecosystems in terms of how we perceive and manage their 'naturalness'. Some terrestrial habitats considered to be of high conservation value are semi-natural, in that positive intervention through the maintenance of certain 'traditional' human activities, such as grazing by livestock, controlled burning and small-scale horticulture, tends to increase their structural and trophic complexity by maintaining habitats in different forms and at different successional stages, therefore human intervention is considered to be required to conserve them in their modified state. Of course, many terrestrial habitats have been subject to negative interventions through the homogenising impacts of human activities such as intensive agriculture and forestry. The focus for managing terrestrial protected areas is therefore to restrict such activities, but also to promote certain human interventions.

By contrast, marine ecosystems are generally natural in management terms, in that they are rarely considered to be valuable as a result of positive intervention by humans. Humans have, of course, intervened in many if not most marine ecosystems, but these interventions are usually considered to be negative. For example, the impacts of fishing tend to reduce the structural complexity of marine ecosystems, thereby 'homogenising' them and reducing their biodiversity (Thrush et al., 2006; Watling and Norse, 1998). Fishing also tends to reduce the trophic (Pauly et al., 1998) and functional (Micheli and Halpern, 2005) complexity of marine ecosystems.

This leads to significantly modified ecosystems and the majority of the world's coastal seas have been affected. However, it is rarely argued that such activities should continue in certain marine areas because the impacted habitats are considered, as a result, to have developed a conservation interest. The general approach to the management of MPAs is therefore one of non-intervention in comparison to the active management approach to conservation which is often practised on land. MPA management essentially involves the minimisation of negative interventions, through the restriction of certain activities in certain areas, in order to maintain relatively natural ecosystems, rather than the promotion of positive interventions through the selective continuation of certain activities, in order to maintain semi-natural habitats. This can be perceived as an exclusionary, 'humans-out' approach to governance that can present challenges when trying to promote cooperation through governance initiatives.

Property rights

Land tends to be owned by specific parties such as individual people or groups of people, commercial companies (private property rights) or the government (state property rights). Some terrestrial areas are subject to community property rights, whereby local people have the rights to live on and use the land, or common property rights, whereby people have rights of access to undertake certain extractive and non-extractive uses under customary and/or legal common property rights, but most land is subject to private or state property rights. Land is often also leased or rented to certain people for certain uses and can be sold from one party to another. Whilst the owners or occupiers of land may not have complete legal autonomy over how the land is managed, they generally do have the rights to determine who can use the land and in what ways such land can be used. Land owners may be subject to state controlled restrictions for biodiversity conservation purposes within terrestrial protected areas, but land areas are generally subject to relatively well-defined rights of access and use. This makes it relatively straightforward for such restrictions to be implemented, as the state can define the owners, leasees and people with rights to use a given land area, whose activities it needs to regulate, with relative ease.

Marine areas and resources under national jurisdiction, on the other hand, have relatively poorly defined rights of access and use, and are rarely subject to private property rights, such areas and resources generally being recognised as state property. State ownership of marine areas and resources within the Exclusive Economic Zones (EEZ) has recently been legally specified under the United Nations Convention on the Law of the Sea (UNCLOS, 1982), but the reality is that marine areas and resources are subject to complex combinations of state, open access (often *de facto* where state capacity to regulate is lacking), community, and common property (often having evolved in seas subject to *de facto* open access) regimes, private property regimes being a rarity in our seas, hence the customary principle of the 'freedom of the seas' that UNCLOS both reinforces and challenges. Some users consider the marine realm to be one of the 'last of the commons' and are culturally resistant to restrictions on freedom of access, compared to terrestrial areas, where private property rights and restricted rights of access and use are *relatively* well accepted (Jones, 2001). This resistance may be partly attributed to the tendency for such restrictions to be imposed by the state, rather than by private property right owners. It could be argued, however, that the state is merely imposing restrictions in its capacity as the authority responsible for fulfilling strategic societal objectives concerning our seas, given that UNCLOS assigns such responsibilities to the sovereign state for seas under national jurisdiction, as is discussed further, in terms of positionality, in Chapter 5.

Marine areas arguably also tend to be able to support a wider diversity of uses than terrestrial areas, due to their three-dimensional nature coupled with customary expectations concerning the freedom of the seas and assumptions about the resilience and productivity of our seas. This is further complicated by the sectoral basis on which different marine activities tend to be regulated amongst different authorities, leading to 'turf battles', regulatory gaps and confusion amongst different sectoral agencies and their policies. These tendencies and the related complexity of property rights regimes leads to 'multiple uses amongst multiple users' regimes that pose major challenges for MPA governance, recognising that MPA designations themselves alter the marine property rights regime of the sea area in question (Mascia and Claus, 2009).

There is currently considerable interest in the potential of assigning private property rights to both fisheries and particular marine areas, as a means of improving governance and achieving certain conservation objectives. Where such property rights can be defined, there may be significant potential in what represents a novel approach in the marine environment but a routinely employed approach in the terrestrial environment, as is discussed in the next chapter. However, in general the relative lack of defined marine property rights, the perception of our seas as one of the last of the commons, the tendency for multiple uses amongst multiple users, and the related confusions and conflicts amongst sectoral regulatory frameworks, pose major challenges for the governance of MPAs.

Summary

This chapter has discussed some of the key differences and divergences that underpin many MPA governance challenges, particularly different value priorities and related divergences amongst marine ecologists and fisheries scientists, as well as some key differences between marine and terrestrial ecosystems. Roberts (2012b) cites fisheries biologist Marcel Herubel's lament dating from 1912 that 'the exigencies of theory often accord ill with corporate interests, and the multiplication of coastal reserves would quickly arouse the anger of fishers'. The fact that this is still the case after one hundred years and that we still have so few no-take MPAs, despite growing scientific and societal concerns about the state of our seas, is arguably largely attributable to such differences and divergences. Against this background, the next chapter will discuss key theories on how some of the challenges related to these concerns might be addressed through different governance approaches.

Chapter 4

Different theoretical perspectives on governance

Introduction

Governance can be defined as 'the involvement of a wide range of institutions and actors in the production of policy outcomes ... involving coordination through networks and partnerships' (Johnston et al., 2000, 317). Debates about how to govern MPAs, along with their terrestrial counterparts, are taking place in the much wider context of debates about how we should go about governing people and the social, economic, political and bureaucratic systems of which they are a part. These debates are not confined to recent times, Plato long ago (*The Republic*, 360BC) having discussed the role of the state in 'steering' human affairs, the word 'governance' being derived from his use of the Greek verb 'to steer'. Since Plato, many other seminal thinkers, such as Machiavelli, Hobbes, Smith, Marx, Weber and Hardin have put forward various observations, ideals and theories concerning the relative importance of the roles of states, markets and civil society as sources of governance steer. A full consideration of these debates about where governance steer should come from is beyond the scope of this book, but they can, in essence, be considered in terms of three perspectives (Table 8).

Many social and political scientists have taken these debates forward through studies of natural resource governance issues. Ostrom (2007), for example, highlights the importance of providing for systematic analyses of the attributes of different governance case studies, recognising their linkages with wider socio-economic, political and ecological structures. Many governance analyses, however, remain primarily focused on the role of people and civil society in self-organising social-ecological systems e.g. Ostrom (2009), and resistant to the potential role of some degree and form of state control and regulation. This is consistent with Kjær's (2004) observation that whilst governance analyses should consider the role of the government and state steer, since the 1980s governance has increasingly been considered by many analysts as being distinct from government in its focus on people and civil society.

This book will take a more rounded view of governance, which is here defined as 'steering human behaviour through *combinations* of state, market and civil society approaches in order to achieve strategic objectives'. This definition is consistent

Table 8 Three perspectives on sources of governance steer (discourses after Dryzek, 2013)

Steer type	Decisions taken by	Characteristics	Discourse
State steer	Governments and regulatory agencies	Top-down decisions by state through laws and regulations, drawing on expert advice	Administrative rationalism: 'leave it to the experts'
Market steer	Markets and economic systems	Decisions on basis of economic rationality through markets and/ or implemented through economic incentives, including property rights	Economic rationalism: 'leave it to the markets'
Civil society steer	People, social networks and related organisations	Bottom-up decisions through deliberations amongst individuals, community/non-governmental organisations and social/family networks	Democratic pragmatism: 'leave it to the people'

with the growing recognition in governance debates that there is a need to move beyond ideological arguments as to which governance approach is 'right' or 'best' and, instead, develop governance models, frameworks and approaches that combine the 'steering' role of states, markets and people. Such integrated, pragmatic perspectives should enable us to move on from debates about whether we should rely on the strong hand of state power, the 'invisible hand' of market forces or the democratic hands of the people, and to consider how these three approaches can be effectively combined. This more rounded view is consistent, for instance, with Kooiman et al.'s (2008) focus on interactions between civil society, state and private actors in governance.

Current environmental governance theories

Introduction to neo-institutionalism

This book draws and builds on recent theories concerning the governance of natural resources from a neo-institutional perspective, i.e. a sociological perspective, rather than an economic one, on how institutions evolve and can shape and be shaped by the behaviour of people. In particular, it draws and builds on several inter-linked neo-institutional theories that relate to promoting sustainable social-ecological systems, as outlined in Table 9. These are premised on a critical view of hierarchical systems, whereby the government authority at the highest level of the bureaucratic hierarchy exerts top-down control of society via the lower levels of the hierarchy (Figure 5a). Instead, they focus on how institutions can evolve to promote sustainable communities, taking a very broad view of institutions – 'Prescriptions that humans use to organise all forms of repetitive and structured interactions,

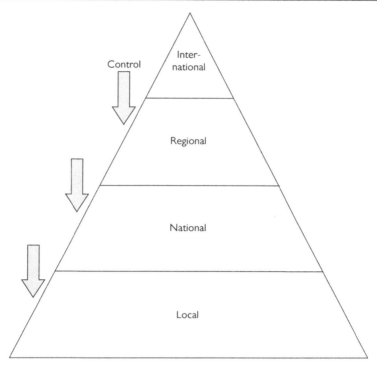

Figure 5a Key governance concept: Top-down 'command and control' hierarchical governance.

including those within families, neighborhoods, markets, firms, sports leagues, churches, private associations, and governments at all scales' (Ostrom, 1995, 3).

Based on very many case studies over many years, neo-institutionalist researchers have observed that top-down hierarchical governance approaches (Figure 5a) tend to be ineffective in achieving their aims. By combining empirical findings of case studies involving other governance approaches with theoretical perspectives and experimental learnings, they have developed an alternative concept – bottom-up, polycentric systems (Figure 5b). Such systems are bottom-up, in that the emphasis is on self-governance by local people. They are polycentric, in that the focus for reaching and implementing decisions is based at a local level in various centres or 'places'. Much of the current thinking and research on natural resource governance is focused on the idea, or perhaps the ideal, that such place-based self-governance represents the most effective way forward for equitably addressing the challenges of sustainably governing natural resources. It is therefore worth discussing these concepts before going on to discuss some related challenges.

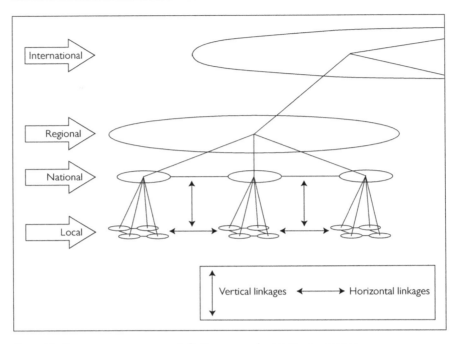

Figure 5b Key governance concept: Bottom-up polycentric governance.

Place-based governance

There is an increasing focus and emphasis on promoting the development of institutions to govern marine resources at a local scale. Such 'place-based' approaches are considered an effective way of governing complex, heterogeneous and dynamic ocean ecosystems by: (a) providing for a focus on spatially bounded individual places that represent manageable small-scale units; (b) providing for the participation of a manageable number of local users in a local governance initiative, thereby promoting the development of social capital, including a sense of shared responsibility and community stewardship amongst local people for 'their' area and natural resources; (c) harnessing the traditional long-term and fine-scale knowledge of local users; and (d) obtaining fine-scale feedback to local management interventions from both natural and social systems in order to reduce uncertainties and provide for management to be adapted in the light of collective learnings from such feedback (Wilson, 2006; Young et al., 2007).

The concept of place-based governance is consistent with two of the design conditions identified by neo-institutional researchers that facilitate the evolution of institutions for the sustainable governance of shared or common-pool resources (CPRs) – small size and well-defined boundaries (Agrawal, 2001). Accordingly, a key assumption on which this concept is based is that marine ecosystems can be spatially

Table 9 Main concepts that relate to promoting sustainable social-ecological systems

Concept	Essence of concept	Key source
Conditions that promote sustainability through building social capital	Conditions that affect the likelihood of self-organisation through building social capital – mutual trust and confidence that co-operation will be reciprocated – and promoting collective action in order to sustainably govern common-pool resources (CPRs)	Ostrom (1990) – Ostrom (2009)
Polycentric governance	'Nested quasi-autonomous decision-making units operating at multiple scales'	Folke et al. (2005) building on McGinnis (2000) and Ostrom (1996)
Social-Ecological Systems	Dynamic inter-linkages and feedbacks in coupled, complex and co-evolutionary social and ecological systems	Berkes and Folke (1998)
Complex Adaptive Systems	Structure and function of social and human systems respond to internal and external drivers in an adaptive cycle that can lead to transformations, whereby the systems may shift to a different state	Gunderson and Holling (2002)
Social-Ecological Resilience	'Capacity of a [social-ecological] system to absorb disturbance and re-organize while undergoing change so as to still retain essentially the same function, structure, identity and feedbacks'. Influenced by the adaptive and transformative capacity of the actors in the social system	Walker et al. (2004)
Adaptive Collaborative Management	How self-organised systems can learn by experience and adapt in the face of crises, as a result of collaborations and cooperation amongst actors at multiple institutional levels and spatial scales through networks, in order to promote collective learning and collective action, potentially transforming the social-ecological system to a more sustainable trajectory and resilient state.	Folke et al. (2005)

sub-divided into local sub-systems or small sized 'places' that are 'almost', 'somewhat' or 'nearly' independent (Wilson, 2006), in that they have limited social and ecological interconnections with neighbouring places.

One of the key challenges that place-based governance has to address is that, in reality, marine ecosystems can rarely be broken down into discrete areas that can be governed independently, as different ecosystem components vary from micro to mega in scale, therefore they function at multiple spatial scales, complexity and uncertainty tending to increase with increasing spatial scale (Olsen et al., 2011). This means that whilst some populations and the ecological structures and processes that support them may operate at a local scale, bound within a given place, other populations and the structures and processes that support them may operate at wider spatial scales that operate beyond the bounds of a given place. The relatively wide scale of some ecosystem components can be explained by recognising that there are usually ecological linkages and connections between different marine places, through factors such as the movements and migrations of fish across their range, the drift of plankton, nutrients, etc. in currents, and source-sink connections between spawning areas and larval settlement areas. Whilst recent findings are challenging previous assumptions about the distance across which such interconnections occur, given that some sub-populations have relatively narrow distributions (Kinlan and Gaines, 2003; Hauser and Carvalho, 2008), the scale and connectivity of marine ecosystems tends to be across wider scales than for terrestrial ecosystems, presenting significant challenges for place-based management.

Furthermore, given the multiple-users of marine areas and their increasing mobility, e.g. fishing vessels that tend to move considerable distances on a legally allowed, unregulated or illegal basis between different grounds, there also tend to be human interconnections between different places through the movements and migrations of people. With respect to both ecological and human interconnections, governance through place-based management has to address the same challenges as MPAs, which are discussed in Chapter 3 in terms of scale and connectivity, as MPAs are essentially 'places'. Given these interconnections, a narrower focus on self-governance through place-based management is rarely appropriate for all ecological and human components.

There is, as is further discussed later in this chapter, recognition amongst neo-institutionalists of the need for linkages between places and with higher level institutions in order to address human and ecological interconnections (e.g. Cash et al., 2006), but their governance analyses remain primarily focused on place-based self-governance. For instance, whilst Brondizio et al. (2009) recognise that 'place-based management is sensitive to outside forces and involves a concerted effort to view places as complex and dynamic systems that are open rather than closed in character', the self-governance premise of their analyses is revealed by their particular focus on approaches such as 'building vertical linkages from the bottom-up' and the 'self-organizing processes that social capital facilitates'.

Self-governance

The focus on promoting self-governance is consistent with and arguably stems from the wider participative development focus on empowering local people, in order to promote locally appropriate development, rather than imposing modern Western models of development (Chambers, 1983). Self-governance is also arguably the concept on which place-based governance is premised, as the 'place' is essentially an appropriate context for self-governance. It is also a key element in the concept of polycentric governance, of which many other governance concepts are elaborations, each local 'centre' being defined as quasi-autonomous (Table 9), i.e. having some degree of independent decision-making power without interference from outside authorities. This focus on quasi-autonomy is derived from one of the original design principles for the sustainable governance of CPRs – that 'the rights of appropriators [users] to devise their own institutions are not challenged by external governmental authorities' (Ostrom, 1990, 101) – which is itself consistent with the focus of participative development on empowering local people. Accordingly, the state's role is one of allowing considerable local autonomy whilst providing a supportive framework for and facilitating the development of self-organised local governance institutions, rather than one of exerting control over local governance. This principle has been incorporated into Agrawal's (2001) synthesis of design conditions for the evolution of institutions for sustainable natural resource governance, two of which are 'locally devised access and management rules' and 'central governments should not undermine local authority'.

The focus on self-governance is arguably strongly influenced by the 'Habermasian' concept of communicative rationality, i.e. people can reach a common understanding through reasoned arguments and deliberations, and thereby agree coordinated actions based on consensus and cooperation (Habermas, 1984, 86). Whilst self-governance is not necessarily seen as a panacea (Ostrom, 2007), the condition that central governments should not undermine local authority is still considered to be one of three basic necessary conditions for the sustainable governance of CPRs (Stern et al., 2002). Recognition of the necessity of self-governance is still pervasive in many contemporary governance analysis concepts, e.g. the focus on self-organised systems in adaptive collaborative management or quasi-autonomous systems in polycentric governance (Table 9). It could thereby be argued that many governance analyses remain constrained by an underlying commitment to Habermasian influenced ideals, i.e. that the state should facilitate and support deliberations amongst local people, and assist in enforcing decisions, but it should not interfere with or undermine such deliberations (Jones, 2013a).

Scale challenges

Discussions related to place-based self-governance have long recognised that such 'places' cannot be considered in isolation, one of Ostrom's (1990, 101) design principles having emphasised the need for 'nested' approaches, whereby the governance activities of smaller organisations are integrated with those of larger organisations, of which they are a part. This nested, multiple-layered approach recognises that such larger organisations generally have a wider spatial remit, including organisations that represent other users of the CPR in question in neighbouring and nearby places, as well as local, regional and national government authorities. Similarly, McCay (2002) recognises that local people in a given place are 'embedded' in institutional structures that operate at local and wider scales, for example, wider policy frameworks and markets, and that the influence of these wider structures on the evolution of local governance institutions in a given place needs to be integrated into CPR governance analyses. This recognition that people and institutions in a given place are nested or embedded in wider-scale institutions is reinforced by growing recognition of the human and ecological interconnections discussed above in relation to place-based management.

There is much discussion of the nature and influences of globalisation, and whilst there are disagreements on whether this is a phenomenon or a narrative, whether it began millennia, centuries or decades ago, etc., there is a common recognition that human interconnections are increasing in terms of their scale and influence, be this through the increasing scale of global markets, the increasing mobility of people, the increasing influence of media such as television and the internet, etc. Accepting that we live in an increasingly globalised world, in which the scale and influence of human interconnections is increasing, there is growing recognition that research into natural resource governance issues needs to move on from its focus on case studies in particular places that are characterised by self-governance amongst local people of spatially bound natural resources (Agrawal, 2001; Stern et al., 2002; Jones and Burgess, 2005; Berkes, 2007). This recognition is commonly discussed in terms of the scale challenges that these human interconnections raise for natural resource governance (Cash et al., 2006).

As well as these human interconnections, the importance of ecological interconnections was also discussed above as a challenge for place-based self-governance, in that ecosystems, like human systems, operate at different spatial scales. Whilst proponents of place-based management tend to focus on spatially bound fine-scale examples of ecosystem structures and functions, the reality is that ecosystems operate at a variety of spatial scales, from micro to mega (Olsen et al., 2011), and that there are ecological interconnections between different places. It is important to recognise that such ecological interconnections also raise scale challenges for the governance of natural resources in a given place, as account must be taken of the reality that human interactions with an ecosystem in one place can have consequences in an ecosystem in a spatially distant place, and that the governance of human interactions with marine ecosystems within a place cannot address the impacts of interactions that occur in other places.

The dimensions of scale challenges are illustrated by Figure 5b, the lowest level of which represents local places at which governance decisions must ultimately always be implemented and at which local users should be involved in such decisions. Whilst the next level up in Figure 5b represents a national level, the reality is that human and ecological interconnections will occur at a variety of scales, from micro to mega, representing a hierarchy. Similarly, decisions will be taken at a variety of scales, some at a local level, some at higher level between local and national, some at a national level, etc. The concept of social-ecological systems (Table 9) is premised on this recognition that human and ecological systems are interconnected or inter-linked at different scales in what might be considered as coupled hierarchical systems. In the context of social-ecological systems, neo-institutional researchers recognise that environmental governance outcomes are likely to be negotiated amongst actors at different spatial scales and at different institutional levels (Cash et al., 2006), horizontal and vertical linkages (Figure 5b) representing the channels for such negotiations (Berkes, 2002).

A critical set of questions remains, however, given the human and ecological interconnections between different places, and thereby the potential for disagreements and competition for resources between users of different places – how will such negotiations be coordinated, who will coordinate them, how will disagreements and conflicts be resolved and who will take the final decision in the face of unresolved conflicts? More importantly, how can the potential need for coordination and conflict resolution between governance initiatives in different places be reconciled with the self-governance principle? This principle is arguably integral to all of the concepts summarised in Table 9, all of which are based on the assumption that local decision-making in each place will have some degree of independence, and that higher level state institutions will confine their role to facilitating and supporting local self-organised governance institutions, rather than undermining and attempting to control them.

This principle is illustrated by the legend to a figure depicting polycentric governance in the Caribbean (Fanning et al., 2007, similar to Figure 5b), which states that the 'multi-level linkages do not necessarily imply a controlling function'. Instead, Fanning et al. (2007) consider that 'a network structure where the linkages facilitate self-organization, which may be negative or positive relative to the goal of sustainability' is more likely in large marine ecosystems such as the Caribbean, which have numerous levels of geographic and jurisdictional scales. Whilst their recognition that self-governance may lead to potentially unsustainable decisions and actions is refreshingly frank, it is also unusual, in that many neo-institutional researchers seem to have a tremendous faith in the Habermasian ideal that facilitated deliberations and negotiations will lead to consensus and cooperation amongst local people in a given place that yields sustainable outcomes. Whilst the state is recognised as having a role in providing a framework for and facilitating self-governance, including establishing general rules at a large system level 'that can be tailored to local circumstances' (Ostrom, 1996) and providing some degree of coordination and even influence, the principle that the state will not interfere with or attempt to control place-based self-governance remains both fundamental and pervasive.

This is the key point of departure of this book from neo-institutional concepts of natural resource governance, in that whilst they recognise the nested and multiple-layered nature of institutional structures and the scale challenges that are raised by the human and ecological interconnections, or 'functional inter-dependencies' (Brondizio et al., 2009) between different places in social-ecological systems, they remain adherent to the principle of place-based self-governance and convinced that deliberations and negotiations within and between places, via 'seemingly endless possibilities for combinations of horizontal and vertical linkages' (Berkes, 2011), will address such challenges. Neo-institutionalists recognise that empirical evidence for the effectiveness of multiple-layered governance in addressing scale challenges is lacking (Newig and Fritsch, 2009), and that the role of the state has received little attention (Duit, 2012), even though some case studies demonstrate that the state plays a critical role in managing resilience (Armitage and Plummer, 2010), but the primary emphasis of much neo-institutional research on the theme of social-ecological resilience remains on place-based self-governance and resistant to constructive recognition of the role of state control.

Against this background, it is argued that some scale challenges require some form and degree of state control in order to effectively and equitably address them, as facilitated negotiations alone will often not be sufficient to address such challenges. The particular scale challenges that are raised by MPAs will be considered later in this chapter, as will the related concept of MPA co-management, but it is first worth considering an alternative concept of governance in the light of this departure from the neo-institutional concepts and their presumption against state control.

Alternative concept of co-evolutionary hierarchical governance

In the shadow of hierarchy

There has been a long-standing recognition of the scale challenges inherent in natural resource governance. Dryzek (1987, 96–100) discussed the coordination difficulties inherent in hierarchical approaches, in that higher levels may have a wider-scale and more strategic perspective on ecological problems, and lower levels may have a better conception of the complexities of the 'spatially bounded' local-scale at which ecological problems must ultimately be addressed, but that it is very difficult to decompose such problems to an appropriate level of detail to be addressed at local-scale levels, whilst also ensuring that the different levels will act in concert to strategically address wider-scale challenges. He subsequently discusses this as a root problem of environmental governance, i.e. that problems of any complexity defy centralisation but that decentralisation can undermine the integration required to strategically address wider-scale, interconnected challenges (Dryzek, 2013: 93–94). He rejects inflexible 'administered hierarchies' ('strong thumbs, no fingers') as a means of addressing this dilemma and instead proposes a negotiated compliance or 'rolling rule regime' approach, whereby central agencies set standards, and compliance is negotiated locally on a 'learning by doing' adaptive management basis, with such institutional learnings being transferred through more flexible and ecologically rational hierarchies ('green and nimble fingers') for application in other contexts and places (Dryzek, 1987, 108–109; 2013, 93–97).

This alternative concept of co-evolutionary hierarchical governance (Figure 5c) is more progressive, in that whilst it rejects the 'command-and-control' approach inherent in top-down hierarchies, through which a 'monocentric' state attempts to directly control users of natural resources (Figure 5a), it also accepts that some degree and form of state coordination and control is necessary to address the central dilemma of scale challenges. Dryzek (2013, 97) discusses this as a diluted form of administrative rationalism (i.e. state steer), more linked to democratic pragmatism (i.e. civil society steer) (Table 8), which represents a shift from centralised government to decentralised, informal and networked governance, in a manner that is more consistent with the definition this chapter opened with.

This does not, however, mean that public authority is abdicated, and the associated risks that unrepresentative but powerful private interests will capture local decision-making processes, as the state holds locals to account as to whether centrally set standards have been complied with (Dryzek, 2013, 97), i.e. the state retains some degree of control rather than confining its role solely to facilitation.

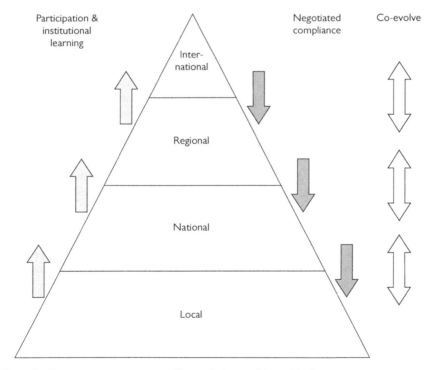

Figure 5c Key governance concept: Co-evolutionary hierarchical governance.

That is why this is an alternative concept of governance to that of the neo-institutional governance concept, which is premised on self-governance, with the state's role confined to facilitation. In this alternative concept, top-down and bottom-up approaches are combined.

Whilst the means of ensuring compliance is subject to negotiation, including the participation of local users and the potential for learnings at higher institutional levels, the obligation to comply with centrally set standards, in order to achieve wider-scale strategic objectives, is not. That is not to say, however, that vertical linkages are confined to top-down obligations to comply, as such negotiations could lead to revisions to the nature and degree of the standards and the related obligations, provided this does not undermine wider-scale strategic objectives, as well as to mutual learnings on the means of complying with them. This recognition that the 'structure' of higher level, wider-scale institutions (including state institutions) and the 'agency' of local people (including private interests) for self-governance can co-evolve is consistent with the environmental sociology theory of structuration (Giddens, 1984). It is also consistent with McCay's (2002) recognition that whilst people are influenced by the structures in which they are embedded, they can also influence and alter these structures.

This alternative concept of co-evolutionary hierarchical governance is a key theme in this book, i.e. the state sets the standards necessary to fulfil strategic societal objectives, but decentralises the authority to achieve these standards to more local levels, subject to accountability to the state that the strategic objectives are being fulfilled. It represents a way of combining state steer and civil society steer (Table 8), recognising that state 'structures' and civil society 'agency' can co-evolve (Figure 5c). As is discussed later in this book, market steer can be harnessed by the state to provide a form of indirect state steer through economic incentives. This more rounded concept of governance, involving *combinations* of all three forms of governance steer, is key to this alternative concept of hierarchical governance. That is not to say, however, that local people in a given place are mere subjects of state control, as they can influence higher level institutions, including having a democratic role in establishing strategic societal objectives and related obligations. The state does not confine its role *solely* to being a passive facilitator of bottom-up approaches *or* an executive controller of top-down approaches.

There is a tendency amongst some governance analysts to assume that the two approaches are diametrically opposite and that any questioning of the principle that the state's role should be confined to the facilitation of bottom-up place-based self-governance and negotiations through linkages (Figure 5b) implies a return to a top-down 'command-and-control' government approach (Figure 5a). This book, however, is premised on the argument that top-down and bottom-up approaches can be combined, along with market approaches (Figure 5c). Accordingly, a key focus for governance should still be to focus on specific 'places', e.g. MPAs, but to also focus on the potential of this alternative concept of hierarchical governance, whereby the state decentralises rather than relinquishes authority, providing for the co-evolution of top-down and bottom-up approaches and the fulfilment of wider-scale strategic objectives (Figure 5d). This view of governance resonates with some elements of Rip's (2006) discussions on a co-evolutionary approach to reflexive governance, though these are more focused on self-governance amongst networks of actors.

This co-evolutionary hierarchical governance concept is more consistent with arguments that the state is still a key actor that actually continues to play a crucial role in governance (Bell and Hindmoor, 2009), contrary to the view amongst many governance analysts that government by the state has been replaced by networked governance amongst civil society actors and organisations. Rather than retreating, or having been hollowed out or replaced by networks, states are considered to have actually been repositioned and/or reconfigured, and continue to provide a regulatory steer in order to achieve strategic societal objectives that only the state is in a position to achieve. Rather than primarily attempting regulatory steer through direct command and control, states have adapted their roles to provide governance steer through persuasion, partnerships, markets, communities and associations, as well as the potential to resort to direct regulation. However, the state relies primarily on indirect control by various means, such as attaching conditions to property rights, setting standards that organisations with decentralised authority must comply with, etc., ultimately retaining the right to withdraw such property rights and any decentralised authority associated with them, and revert to direct command

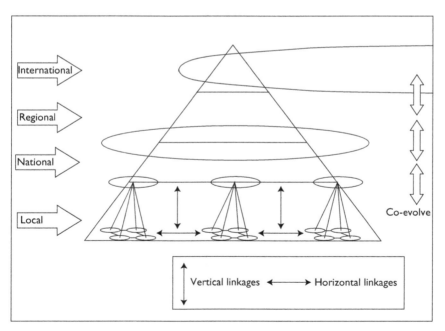

Figure 5d Key governance concept: Polycentric view of co-evolutionary hierarchical governance.

and control, i.e. state steer is achieved *through* coalitions with various non-state actors 'in the shadow of hierarchy' (Bell and Hindmoor, 2009, 71–114). This book builds on this alternative concept of co-evolutionary hierarchical governance, whereby the state retains a degree of hierarchical control over steer in order to ensure that strategic societal objectives are achieved, but different approaches to governance, including through partnerships with local people and private interests, can still be employed to help achieve such objectives in different 'places', Figure 5d illustrating a polycentric view of the co-evolutionary hierarchical governance concept.

Are hierarchies necessary to address scale challenges posed by MPAs?

As was noted above, MPAs essentially represent 'places' in which governance initiatives and analyses can be focused, but given the scale and connectivity of marine ecosystems, as discussed in Chapter 3, coupled with the high mobility of many marine users, e.g. interloping fishing vessels, MPAs are places that are subject to relatively high degrees and extents of human and ecological interconnections. This means that MPA governance must be considered at a wider spatial scale, as (to paraphrase John Donne) 'no MPA is an island', in that human and ecological interconnections with the wider marine area affect the governance of that MPA. Furthermore, as was discussed in Chapter 2, individual MPAs are increasingly considered as inter-connected constituents of wider-scale representative and coherent MPA networks. One of the neglected consequences of 'scaling-up' to wider-scale networks of MPAs is that the governance of each constituent MPA increasingly needs to be considered in terms of the strategic societal objectives that the network is aimed at fulfilling and of the human and ecological interconnections amongst MPAs in the network.

A key consequence of such scaling-up from a governance perspective is that decisions concerning each MPA must be taken with regards to its interconnections with other MPAs and its contribution to achieving the wider-scale and strategic societal objectives of the MPA network as a whole, such interconnections often being identified through wide-scale studies by scientific 'experts'. This further reduces the scope for focusing on the self-governance of MPAs as quasi-autonomous places, as their interactions with and contribution to inter-connected MPAs in the wider network, as well as to other marine places in the wider sea, must be taken into account. This implies that there must be a greater degree of coordination of the governance of MPAs that constitute a network, as only higher institutional levels will have a wider network perspective. At present, the guidance on designing MPA networks (IUCN-WCPA, 2008) recommends that both top-down and bottom-up approaches be used, though each approach is considered in its own right on different pages, and the critical question of whether MPAs that constitute a network will require more top-down coordination is not addressed. It is argued here that more coordination through top-down governance elements will be required for

MPAs that constitute a network, but in keeping with the co-evolutionary hierarchical governance concept (Figure 5d) described in the previous section.

The scale and connectivity of marine ecosystems means that MPAs have to be governed in a manner that takes account of human and ecological interactions at a wider spatial scale, especially as MPAs are scaled-up to networks. This means that higher institutional levels must be engaged in MPA governance, as only these higher levels have an appropriately wide spatial and inter-sectoral remit. This presents significant challenges for the governance of a MPA in a given place, as the need to coordinate governance activities across a wider spatial scale means that the role of higher level institutions must extend beyond facilitating place-based self-governance, towards coordinating and even controlling activities in a given MPA. That is not to say, however, that MPAs become passive subjects of top-down 'command-and-control' from higher level institutions, as there will still be the potential for the degree and means of compliance with MPA management restrictions to be negotiated with local users. There will also be an important related role for the participation of local users and the potential of such participation to influence higher level institutions, including negotiations on the nature of management restrictions that users of a given MPA would be expected to achieve compliance with, as in the concept of co-evolutionary hierarchical governance illustrated in Figure 5d. Furthermore, many MPAs will rely on such higher level institutions to address human interconnections, e.g. to restrict incoming fishing vessels that could otherwise undermine the fulfilment of local conservation and sustainable use objectives, and ecological interactions, e.g. to restrict fishing in other places that share the same fish stocks as the MPAs.

Addressing scale challenges thereby represents a situation of both co-dependence, whereby local and higher level institutions rely on each other, and co-evolution, whereby local and higher level institutions affect the development of each other. This is a way of representing and elaborating on the more straight-forward statement that the 'design and management of MPAs must be both top-down and bottom-up' (Kelleher, 1999, xiii), specifically in the context of wider concepts of governance and related debates. Given the particular scale challenges that MPAs pose, it is argued that the co-evolutionary hierarchical governance concept is more appropriate and realistic than adherence to place-based self-governance.

Governance challenges related to divergent objectives

What is it that differentiates the governance of a MPA from the governance of wider seas? The simple answer to this question is that certain conservation/restoration-focused objectives are prioritised by virtue of the MPA being designated as 'protected'. The nine main categories of objectives which MPAs are aimed at achieving were discussed in detail in Chapter 2, as was the tendency that differences between MPA objectives and the objectives of users of a given area are at the root of many MPA governance challenges. The governance challenges related to such differences in objectives are often premised on different value priorities, as was discussed in Chapter 3. These differences can raise basic conflicts, MPAs often being more focused on objectives related to conservationist values, MPA users often being more focused on objectives related to utilitarian values. Such basic conflicts can be particularly challenging to resolve as they are often based on different ethical perspectives, compromises between which may be very difficult to reach.

Given the major governance challenges of addressing such basic conflicts, it is tempting to try to avoid them by various means. One way of doing this would be for the objectives of a given MPA to be negotiated at the very outset of the process, before the MPA is actually implemented. This very early stage has been discussed as representing 'step zero', which is considered to be critically important, as any flaws in these early decisions, including the ways in which they are reached and the actors involved in reaching them, could taint and undermine subsequent governance processes. In particular, imposing MPAs in a way that is premised on certain *a priori* objectives put forward by governments, conservation biologists and NGOs could be interpreted as an attempt to exert state control, particularly where the values of local users may differ to those of the state or other non-local actors. Instead, step zero should include a process by which objectives are empirically investigated, so that the views and agendas of local users can be included in the process of establishing the objectives of the MPA (Jentoft et al., 2011; Chuenpagdee et al., 2013), rather than attempting to impose pre-determined objectives in a top-down manner.

Such an approach resonates with arguments that conservation should focus 'on participation as a process, in which the objectives and actions are not settled in advance but emerge from the act of participation itself' (Goodwin, 1999). However, extending bottom-up approaches to decisions on what the objectives of MPAs should be, or, indeed, whether a MPA should even be designated, could lead to wider-scale strategic societal objectives related to conservationist values, including those related to MPA targets (Table 1), being undermined. The priorities of local users may differ to or even conflict with the strategic conservation objectives of higher level state institutions, leading to a fragmentation and even failure to achieve such objectives (Goodwin, 1998; 1999; Pennington and Rydin, 2000).

Stern et al. (2002) also recognise that the vertical linkage of global conservation objectives with local objectives raises extreme governance challenges as objectives often differ at different institutional levels. This is consistent with arguments that local users tend to be more focused on (hopefully sustainable) resource exploitation whilst national to international state institutions tend to be more focused on wider-scale strategic societal conservation objectives (Jones, 2013a). This is particularly the case for MPAs, given the growing societal and scientific concerns about the declining state of marine ecosystems discussed in Chapter 1, the conservation/restoration focus of the MPA objectives discussed in Chapter 2 and the basic conflicts between utilitarian and conservationist values discussed in Chapter 3. The linkage of local, national, regional and international management institutions to achieve a balance between local and strategic objectives is therefore very challenging to achieve. This challenge is arguably one that has been neglected by the CPR literature in its focus on local institutions and neglect of external contexts (Agrawal, 2001; Berkes, 2002) and is one that Stern et al. (2002) identify as a priority for future studies, particularly given that the majority of CPR governance case studies have been focused on sustainable resource exploitation objectives.

This book is focused on the co-evolutionary hierarchical governance concept (Figure 5d), which includes recognition that some objectives are pre-determined on an *a priori* basis, particularly those intended to fulfil strategic wider-scale societal conservation objectives and related targets (Table 1) and legal obligations. Arguably, it is a focus on fulfilling these conservation objectives that distinguishes marine *protected* areas from wider marine areas where, at best, sustainable exploitation is the focus. It must be accepted that very few marine areas would be considered to be 'protected' were there to be a reliance on the objectives of governance emerging solely on a bottom-up basis amongst the users of a given area, particularly given the driving forces discussed in Chapter 5. This book recognises that whilst there may be considerable scope for some negotiation on where and how certain pre-determined MPA objectives could apply, the challenge lies in reconciling top-down and bottom-up objectives, recognising that this will often involve a degree of imposition from the perspective of some users, and then combining top-down and bottom-up approaches to achieve these objectives through co-evolutionary hierarchical governance.

Co-management

Introduction to the concept of co-management

In a more applied context than the concepts and theories discussed above, recognition has grown over the last two decades that local users, generally referred to as stakeholders in this context, affected by protected areas should be involved in their planning and management to increase the fairness of decisions and promote local ownership and cooperation. This key aspect of the new paradigm for protected areas was recognised at the IUCN's 5th World Parks Congress (Phillips, 2003). The term 'collaborative management' (henceforth 'co-management') describes a range of approaches to provide for the participation of different stakeholders in protected area management (Jones and Burgess, 2005). Such approaches can vary between merely consulting users on proposals and providing an opportunity for their feedback and recommendations on such proposals, through to fully delegating many responsibilities to users. Invariably, co-management builds on the foundation that all stakeholders should participate in decisions in a manner that provides for them to substantially influence such decisions. It is recognised as a 'broad concept spanning a variety of ways by which the agency in charge and other stakeholders develop and implement a management partnership' (Borrini-Feyerabend, 1999, 228). Co-management is thereby an applied expression of the challenges of combining top-down and bottom-up approaches as it is based on partnerships between local users and state agencies in order to achieve protected area conservation objectives. As such, it also illustrates many of the challenges discussed in the previous sections, such as establishing vertical linkages, negotiating protected area objectives, etc.

There has been a tendency in some quarters for co-management to be considered as synonymous with bottom-up community-based management (e.g. Murphree, 1994), whereby the emphasis is on empowering local people in protected areas through reinforcing their proprietorship[9] of these areas, which the intervention of the 'agency in charge' is considered to undermine (Jones, 2013a). This is understandable as co-management emerged from growing recognition that the previous top-down (Figure 5a) 'fortress conservation' approach was often ineffective and nearly always inequitable. One is sometimes inclined to wonder, however, whether the pendulum may have swung too far in the paradigm shift from top-down to bottom-up management, hence the focus of many current governance concepts, particularly the neo-institutional concepts discussed above, on self-governance, with the state's role ideally confined to facilitating decisions amongst local people and assisting in their enforcement.

Co-management as a balancing act

Co-management was intended as a framework for a more balanced approach, in which top-down and bottom-up approaches are combined and a balance is achieved between empowering local people, in order to ensure that their rights, priorities and objectives are provided for, and allowing the state to retain enough power to exert an appropriate degree of steer, in order to provide for the conservation objectives, by which a 'protected' area is arguably distinguished and defined, to be achieved. One way in which this balance can be considered is that it is trying to avoid, on one hand, the risks of imposition, and, on the other, the risks of parochialism (Jones and Burgess, 2005; Jones, 2013a).

The risks of imposition

There are concerns that the very act of designating an area as 'protected' and endeavouring to implement it through co-management represents a continuation of state imposition. Such concerns are based on the potential of co-management to undermine local governance institutions through the top-down imposition of protected areas. The pursuit of co-management through stakeholder participation, under the guise of community-based management, has been criticised as representing a 'new tyranny', perpetuating the injustices of fortress conservation, which was the general approach to protected area governance until the 1970s, by imposing goals and institutions on local people, overriding existing legitimate decision-making processes and displacing other potentially effective governance approaches (Ostrom, 1999; Cooke and Kothari, 2001; Brockington, 2002). Jentoft (2004) argues that MPAs are often pursued because they are perceived to fit well with some pre-conceived, general ideas of what constitutes good governance, but that their inappropriate implementation may lead to resistance and failure. Nichols (1999) argued that this was indeed the case with collaborative MPAs funded by development organisations in Indonesia, which subverted traditional resource management systems and promoted market economics-driven exploitation by external interests, in collaboration with the state and aid NGOs. These concerns are similar to those of political ecologists, discussed in relation to the MPA objective of controlling the impacts of tourism and recreation (Chapter 2), i.e. that MPAs become vehicles for the marginalisation of local users through 'unholy alliances' between corporate tourism interests, the state and NGOs.

Proponents of co-management also have concerns that protected areas pursued as part of a national policy initiative risk being too 'top-down' to provide for the meaningful participation of local stakeholders, given that most protected areas are identified by national decree and the state decides if and how to involve stakeholders (Borrini-Feyerabend, 1999). They may be too top-down if the state assumes a controller role, consistent with the 'bureaucratic impulse to retain central authority' (Murphree, 1994), imposing management structures at too high a level, rather than facilitating their development from the bottom-up (Rydin and Pennington, 2000). There is the prospect that MPAs, even those pursued on a well-meaning collaborative basis, could be interpreted as representing appropriation[9] by the state through the imposition of a regime which undermines local governance institutions (Dietz et al., 2002). Coupled with a lack of state capacity to enforce the MPA regime, this can lead to competitive over-exploitation of the resources of the MPA by local users, where these had previously been managed through local governance institutions. This can also lead to the designation of 'paper parks' or, 'paper MPAs': formally recognised but beyond the state's capacity to effectively manage in the face of resistance and resentment from local users (Murphree, 1994).

The logical extension of these concerns is that protected areas, even those governed through 'co-management', may not represent the best way forward for natural resource conservation and that the focus should be on promoting the development of wider-scale, self-governance regimes, as these are more effective. Protected areas could actually be contributing to the problem of over-exploitation rather than representing a solution, in that they disempower local people and undermine both traditional means of sustainably exploiting nature and the potential for local people to cooperate with conservation restrictions (Hayes and Ostrom, 2005; Hayes, 2006; Ellis and Porter-Bolland, 2008). These arguments are supported by analyses which indicate that forest areas that are not designated as protected are actually better conserved than 'protected' areas, as a result of community-based initiatives and traditional sustainable natural resource use practices. The reliance on protected areas, which essentially represent top-down initiatives, despite the power-sharing claims of co-management, is thereby called into question, as are claims that protected area co-management represents a more bottom-up approach.

The risks of parochialism

There is a risk that a full delegation of authority to local people could lead to the 'tyranny of localism' (Lane and Corbett, 2005), i.e. governance processes being captured by local elites, leading to less advantaged people being marginalised from access to decision-making processes, including losing any benefits and bearing any costs that may arise, potentially reinforcing local inequities (Ostrom, 1999; Schulman and Anderson, 1999; Cooke and Kothari, 2001; Platteau, 2004; Lane and Corbett, 2005). Accordingly, it could be argued that some top-down controls by the state are needed to avoid such elite capture, mediate competing claims (Lane and Corbett, 2005), and ensure that costs and benefits are not distributed in such a way as to reinforce inequities.[10]

At the same time, other analysts of co-management raise the troubling question as to whether local initiatives are necessarily better than centralised ones for achieving conservation objectives (Western and Wright, 1994, 10). As was discussed previously, local users may be more focused on (hopefully sustainable) resource use objectives rather than strategic conservation objectives, so assigning the authority to local users to determine the objectives of a MPA could lead to a focus on resource exploitation objectives. This potential for the reinterpretation and fragmentation of conservation objectives through more bottom-up approaches to co-management could lead to a hollow victory: protected area approaches that successfully promote resource user participation but that result in a 'picked-over and emaciated carcass of biodiversity' (McClanahan, 2004). The focus of such concerns is that resource exploitation and economic development objectives often dominate local decision-making processes, over-riding objectives to protect biodiversity. Walters (2004) and Saunders et al. (2008), for instance, highlight how mangrove conservation projects which are recognised as a success story for community-based conservation are actually undermining biodiversity conservation objectives by the gradual replacement of diverse natural mangroves with less diverse mangrove plantations, through a focus on local resource exploitation interests.

Terborgh (1999) similarly argues that the dominant emphasis on sustainable development, coupled with increased numbers of 'settlers', is leading to the degradation and destruction of allegedly protected areas. He further argues that local–national political and economic vested interests, if not corruption, are often behind the degradation and destruction of biodiversity, and that much stronger top-down controls are therefore required, through an international body, to ensure the fulfilment of strategic conservation objectives and obligations, particularly in LEDCs.

The key challenge that protected areas face is to address the basic conflict between strategic conservation objectives and resource use objectives, recognising that whilst the latter may be sustainable and include traditional approaches that help contribute to the former, this may not necessarily be the case, particularly given the driving forces, that are discussed in the next chapter, that promote exploitation and can undermine conservation. Advocates of more top-down approaches consider that legally binding and enforced obligations, as proposed by Terborgh (1999), are necessary to encourage governments to address such basic conflicts whilst also providing them with the capacity to do so, particularly in relation to protected areas. This arguably requires the state to exercise a degree of control in certain circumstances, particularly where parochial priorities and related corporate and political vested interests risk undermining strategic biodiversity conservation objectives. As such, they also consider that protected area co-management can provide a framework for parochial and vested interests to undermine biodiversity conservation and that the state must therefore retain the legal authority and capacity to exercise top-down control to ensure that designated areas are actually protected.

Such 'back to the barriers' (Hutton et al., 2005) arguments for the need for top-down approaches to governing protected areas in order to overcome the risks of parochialism could be criticised, particularly from a political ecology perspective, as representing a return to 'fortress conservation' through the neo-colonialist imposition of Western wildlife conservation values (Brockington, 2002), or even a resort to ecofascism (Adams 2004, 224). There is an even greater danger, however, that discourses revolving around the risks of imposition and the risks of parochialism will lead to a polarisation of views on whether top-down or bottom-up approaches to the governance of protected areas, including MPAs, should be pursued. Such polarisation could, in turn, perpetuate a pendulum approach, whereby the emphasis swings from one extreme to the other, both over time and between different protected areas. The reality is that community-based and 'back to the barriers' approaches will tend to be used in conjunction in most protected areas, as an exclusive reliance on one or the other approach is unlikely to lead to satisfactory outcomes in either human or ecological terms (Hutton et al., 2005). This brings us back to the key question that runs through this book: how can top-down and bottom-up approaches be combined?

Sharing power

One way in which this question is addressed is to focus on how power can be shared between the state and local users through co-management partnerships (Borrini-Feyerabend et al., 2007). This can also be considered in terms of power relations, Rudd et al. (2003) reviewing arguments that many MPA case studies do not sufficiently take account of, inter alia, power relations between users and the state, whilst Steins and Edwards (1999) have, in a wider natural resource governance context, identified such relations as a priority for further research. Power relations in partnerships between users and the state are also clearly critical to the establishment of vertical linkages and the balanced approach discussed above, taking power to be the capacity to take or influence decisions that are binding on other actors by mandate and/or influence.

This leads to a more progressive definition of co-management – 'power-sharing in the exercise of resource management between a government agency and a community or organization of stakeholders' (Pinkerton, 1992). Power is an interesting but also a slippery concept in the context of protected area co-management. Both perspectives on the risks of co-management discussed above implicitly or explicitly address the question of whether power should be allocated to the state or to local people. The simple answer to this question is, of course, that it should be shared, but the realities of protected area co-management tend to be much more complicated. At the start of this section, co-management was introduced as aiming to achieve a balance between empowering local people, in order to ensure that their rights, priorities and objectives are provided for, and allowing the state to retain enough power to exert an appropriate degree of steer, the state being 'in charge' according to Borrini-Feyerabend's definition (1999, 228).

But can the state *both* empower local people *and* retain enough power to be sufficiently 'in charge' to ensure the fulfilment of strategic, wider-scale conservation objectives? Can power be shared 'in the shadow of hierarchy' i.e. if some powers are devolved to local people through the assignation of proprietorship,[9] but on the condition that certain conservation objectives are achieved, otherwise the state will recall such powers, has power really been shared? This is discussed further in Chapter 5, in terms of decentralisation but suffice it to say, at this stage, that the state may devolve power, but that it should not relinquish it. A perusal of the discussions in this chapter, which are but a brief overview of the natural resource governance literatures, soon reveals that power is, indeed, too important, pervasive, slippery *and* complicated a concept to be addressed solely by a recognition that power should be shared.

It could be folly to attempt to end this brief discussion of such an important concept as power with reference to Foucault, but I will attempt it nonetheless. In keeping with the recognition in co-management circles that 'power and knowledge are impossible to disentangle' (Borrini-Feyerabend et al., 2007, 421), Foucault considered the nature of discourses, relations of power and the constitution of knowledge to be inextricably intertwined. Arguably his most relevant contribution in the context of this discussion is his conceptual analysis of the shift from a top-down form of social control ('sovereign power') to a more diffuse form of self-policing engendered through social surveillance and normative science ('disciplinary power') (Dreyfus and Rabinow, 1982). The implications of this view will be briefly discussed in the next chapter, in terms of positionality and knowledge issues, but suffice it to say at this juncture that many of the references cited in support of MPA objectives in Chapter 2 could be argued to represent normative science, whilst the co-evolutionary hierarchical governance concept could be considered as an example of the shift from sovereign to disciplinary power.

The pervasive nature of market approaches

It is important to recognise that market approaches, including markets as a means of applying steer and implementing policies to achieve strategic objectives, emerge as key elements in most perspectives on co-management. Advocates of more bottom-up approaches often stress the importance of providing for local people to yield the benefits generated by tourism and other compatible economic development opportunities within protected areas. Such approaches may be discussed in terms of integrated conservation and development projects (ICDPs), in which most of the economic benefits from such developments are yielded by local people, who bear many of the lost opportunity costs due to protected area restrictions (Adams et al., 2004), rather than such benefits being yielded mainly by incoming developers. This was discussed in Chapter 2, in the context of the objective of controlling the impacts of tourism and recreation, in terms of preventing the leakage of such benefits. Connecting local people to external markets is also seen as a means of increasing the probability that community-based conservation will be successful, by increasing local economic benefits from sustainable and compatible natural resource exploitation activities (Berkes and Seixas, 2008), as is the assignation of property rights – tenure – to local people (Hayes and Ostrom, 2005), by increasing their ownership of local resources and thereby promoting their role as responsible stewards.

From a more top-down perspective, there is a growing focus on the ecosystems services that our seas provide, coupled with growing societal and scientific concerns, discussed in Chapter 1, that the delivery of such services is being undermined by the degradation of marine ecosystems (UNEP, 2006, 4–6; 2010, 5–6). The reasons for such degradation are increasingly discussed as being due to market and policy failures, the latter being mainly focused on 'perverse subsidies' to the fishing industry. Accordingly, it has been argued that there is an urgent need to 'catalyse ocean finance' to transform the management of marine resources and thereby restore marine ecosystems. Such proposals include scaling up the use of instruments 'that have helped governments put in place clear incentives to all market players to restore and protect coasts and oceans' (GEF/UNDP, 2012, 6). Such instruments include 'blue carbon' payments to promote the use of certain marine habitats as carbon sinks to mitigate climate change, and the role of MPAs in addressing the impacts of overfishing, including instruments to support the achievement of the CBD target (Table 1) for 10 per cent of oceans to be designated and effectively protected as MPAs (GEF/UNDP, 2012). Alongside the MPA proposals, this report also proposes attaching property rights to fisheries through individual transferable quotas (ITQs) as a wider means of addressing overfishing, which resonates with previous recommendations (Arnason et al., 2008).

The rationale behind such proposals is clearly based on advocacy of the role of markets, including neo-liberalist arguments for the removal of policy barriers to economic instruments. The rationale is also, however, clearly based on recognition of the urgency of the need to arrest the degradation of marine ecosystems, including fish stocks, and promote their restoration, through a combination of public policies, stakeholder participation and market instruments. The proposals include major financial investments[11] in MPAs through international mechanisms such as the Global Environment Facility (GEF), which the report argues would deliver benefits many times greater in terms of increased fish stock yields and other ecosystem services (GEF/UNDP, 2012, 10). MPAs are also a means, amongst others, of implementing blue carbon initiatives, as well as other initiatives related to payments for ecosystem services (PESs) (Lau, 2012; Rees et al., 2012), as they can provide an effective institutional framework for conserving marine habitats and populations, and for providing a context for related PESs.

A key principle underlying market approaches, including funding through PESs, is that whilst states 'have the sovereign right to exploit their own natural resources pursuant to their own environmental policies', they also have responsibilities to ensure that such activities 'do not cause damage to the environment of other states or of areas beyond the limits of national jurisdiction' (Article 3, CBD). Whilst this recognition of sovereign rights may undermine the previous 'common heritage of mankind' principle,[12] recognition of this with regards to natural resources now being confined to high seas[1] under UNCLOS (Article 136), it is also increasingly recognised that some ecosystem services that certain habitats and species provide, such as coral reefs, seagrasses and mangroves, are of international importance. This places a particular responsibility on LEDCs due to their richness in such habitats and species, as the need to conserve such biodiversity and thereby maintain internationally important flows of ecosystem services could place a financial burden on these countries. These burdens can be due to the economic development opportunities that are foregone, at least in the short term until, for example, the recovery of marine ecosystems and fish stocks promotes the delivery of economic benefits through increased ecosystem services from MPAs, and due to the need to develop national biodiversity conservation capacity. Some of these ecosystem services benefits may be indirect, thereby providing no direct economic benefit in return for these burdens. Financial incentives, such as the instruments discussed above (GEF/UNDP, 2012), are intended to promote biodiversity conservation, recognising national sovereign rights, by alleviating these economic costs and burdens through financial incentives.

Advocates of bottom-up approaches, however, could critically question whether top-down incentives, such as PESs, are transferred to all local people who are affected by MPA restrictions, along with the other concerns discussed above related to the risks of imposition, and this raises important social justice issues. Similarly, advocates of top-down approaches could raise concerns that assigning property rights to certain areas to certain people could raise the risks of parochialism. The debates as to the merits of top-down and bottom-up approaches to co-management are thus paralleled in debates as to the merits of market approaches. Whilst market approaches can therefore be considered to be pervasive in debates and initiatives related to MPA governance, it is important to recognise that the aims, means and risks of such approaches are interpreted very differently depending on the perspective adopted, i.e. top-down or bottom-up.

Co-management as a framing device

There is a tendency for co-management to be considered as a solution to the challenge of combining top-down and bottom-up approaches to fulfilling natural resource governance objectives, including for MPAs. However, recognising that there are a variety of co-management approaches, it follows that there are also a variety of views on the concept. Drawing on the discussions above, this variety of views can be considered in terms of three main categories (Table 10). It is feasible that these different views could be derived from experience with different protected areas that themselves represent different approaches, but this merely highlights the point that co-management represents a wide range of approaches on which there are a wide range of views.

Co-management might better be considered as a rather loose and flexible framing device for governance debates on the merits of different governance approaches, perhaps even a meta-narrative. Its focus on partnership approaches and power sharing is initially appealing, but the reality is that it tends to obscure issues as it is often less than clear which governance approach is being discussed in a given case, given that a variety of approaches are captured by this term, and which view is being referred to in a given discussion on the validity of such approaches. As such, it is argued that the concept of co-management, in itself, is too loose and flexible a framework for analysing how different governance approaches can be combined to achieve conservation objectives. Its strengths as a flexible framing device for discussions on natural resource governance are its weaknesses as an analytical device.

Co-management of MPAs

The discussions above have varied in their focus, depending on the context of the references in question, from wider development, to natural resources (particularly forests and fisheries), to protected areas, to MPAs, recognising that these are inter-related. It is worth briefly considering how the concept of co-management is applied

Table 10 Categories of views on protected area co-management

Co-management	Rationales
A more bottom-up approach	Reaction to failures of fortress conservation; emphasis on community-based management (CBM) and assigning proprietorship[9] to local users
Too top-down an approach	Raises risks of imposition; return to neo-colonial fortress conservation; participation is a new tyranny that disguises state control; unholy alliances between corporate tourism interests, the state and NGOs; protected areas are a vehicle for such problems and can be less effective than non-protected areas
Too bottom-up an approach	Raises risks of parochialism; elite capture will reinforce inequities; exploitation objectives will over-ride conservation objectives; need strong top-down state control to address basic conflicts and achieve conservation objectives

as a framing device for discussions specifically on MPAs. Co-management emerged as a concept in medicine in the 1960s, in fisheries in the late 1980s and in MPAs in the late 1990s (e.g. Christie and White, 1997). Since then it has been widely adopted as a framing device for discussions on how MPAs can and should be managed. It is often combined with the concept of adaptive management, which has almost certainly been practised as a 'learning by doing' and 'learning by our mistakes' approach by humans for millennia, but which has, since the late 1970s, been discussed as a systematic learning and sequential re-assessment approach to continually improving natural resource governance (Holling, 1978; Walters and Hilborn, 1978). The combined concept of adaptive co-management (ACM) has been adopted by neo-institutionalists in the context of social-ecological systems (Table 9), with a particular emphasis on power sharing and self-organisation, recognising policies as hypotheses and management as experiments to test these hypotheses, particularly given the high degrees of uncertainties associated with such systems (Folke et al., 2005).

The IUCN recommends co-management in its MPAs guidelines, involving the sharing of control through partnerships between government agencies and local communities. These guidelines recognise that a balance needs to be achieved between top-down and bottom-up approaches, recognising that such partnerships range along a continuum between more control by state agencies and more control by local communities. They also recognise that co-management is one model amongst others, such as private sector partnerships, NGO-run MPAs and community-based MPAs, albeit a very attractive option which is proving an ideal way of involving MPA users (Kelleher, 1999, xii, xxii, 29–35, 54). The later guidelines on establishing MPA networks (IUCN-WCPA, 2008) include similar observations and recommendations. It is no coincidence that this book identifies similar options and challenges as the IUCN guidelines, as one of the factors that instigated this research project was a request from the IUCN's Marine Programme to analyse how different MPAs are governed through different governance models in different contexts, and develop further guidance on governing MPAs.

The empirical framework described in the next chapter was designed to build on the analyses behind the IUCN guidelines, though there are some very significant divergences between the two MPA governance analysis frameworks and their underlying rationales, as the discussions in this chapter indicate. In particular, this book does not consider co-management as a discrete model for analysing MPA governance, nor does it consider sharing control or power to be a discrete theme. In essence, though, this book explores the same key question as the IUCN guidelines – how can top-down, bottom-up and market approaches be combined to effectively govern MPAs? I have previously explored this question in terms of 'middle ground' approaches (Jones, 2001) or 'getting the balance right' (Jones et al., 2011) but now consider it more in terms of combining these different approaches.

Many papers, etc.[13] are implicitly or explicitly focused on variations on the basic adaptive co-management approach to achieving MPA objectives (Figure 6). Whilst this is a useful way of illustrating the basic stages of this process and emphasising its

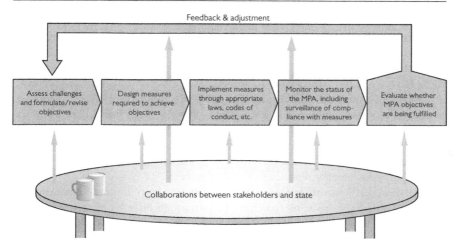

Figure 6 Adaptive co-management process for MPAs.

cyclical 'learning by doing' nature, it actually does little to illustrate how governance challenges are addressed. For instance, whilst stakeholder participation processes are illustrated as having an important input at each stage, the challenges of combining top-down, bottom-up and market approaches, reconciling conflicting objectives between different institutional levels, addressing scale challenges, etc. are not. The appeal of illustrating adaptive co-management as being a simple step-by-step cyclical process by which objectives can be agreed and achieved is understandable, the current guidance on MSP adopting a similar step-by-step approach and illustration (Douvere and Ehler, 2010). The related argument for place-based self-governance through polycentric systems, with conflicting objectives and scale challenges being addressed by negotiations through different combinations of horizontal and vertical linkages, is similarly appealing. However, the appeal of such approaches is also beguiling, in that their apparent simplicity distracts attention from the complicated realities of governance challenges, related to divergent objectives, scale challenges, etc., that will, in reality, have to be addressed. It is argued that such challenges can actually only be addressed 'in the shadow of hierarchy', through the co-evolutionary hierarchical governance concept discussed above.

Social-ecological resilience

MPAs are increasingly considered as elements of coupled, complex and co-evolutionary social-ecological systems. A key theme of related neo-institutional systems theories is that governance should aim to maintain or restore the resilience of such systems (Table 9). Maintaining and restoring resilience is a key aim for MPAs, in that many of their objectives previously discussed in Chapter 2, particularly restoring ecosystems and fish stocks, include aims to promote the resilience of ecosystems and their constituent fish populations. One of the themes in these previous discussions is that depletions or losses of species as a result of over-exploitation can reduce resilience. By preventing or minimising over-exploitation, MPAs have the potential to maintain and increase both the density of populations of a given species and the diversity of different species. This can, in turn, lead to the conservation or restoration of functional group diversity, given that different species have different functional roles in the ecosystem of which they are components. Reductions in species and functional group diversity, and related reductions in functional redundancy and response diversity, are recognised as one of the key factors that reduces ecosystem health and resilience.

'Ecological resilience through species diversity' is a very important concept in relation to this book, as it means that MPAs that are effective in conserving or restoring the density of populations and diversity of species, by preventing or minimising over-exploitation, can confer resilience to the ecosystems covered by the MPA, i.e. the key to ecological resilience is species diversity and MPAs can increase species diversity and thereby resilience. Does this follow for social systems in relation to the role of institutional diversity? This is an important question, recognising that governance can only realistically focus on institutions to steer the actors in the social system, given that ecological systems are steered by ecological interactions and environmental conditions, rather than human interactions. Human activities have the capacity to impact and reduce the complexity of marine ecosystems, as is evidenced by the growing scientific and societal concerns discussed in Chapter 1 about the health of marine ecosystems, but actually steering marine ecosystems by purposefully intervening in their structures and processes is beyond the capacity of humans, as was discussed in relation to the relative naturalness of the marine environment in the previous chapter.

The definition of resilience for social-ecological systems (Table 9) describes what resilient systems have the capacity to do, i.e. absorb disturbance and re-organise, but it does not describe how such capacity is conferred from a social systems perspective. Given that species diversity confers resilience to ecosystems, does it follow that institutional diversity confers resilience to governance frameworks for social systems? Whilst the focus of neo-institutionalists remains predominantly on self-controlling systems in polycentric and multi-layered organisational frameworks, they also recognise that a diversity of institutions at different organisational levels, from individual actors to organisations and governments, can confer resilience to governance systems. In parallel with ecological resilience, institutional diversity is recognised as conferring resilience through institutional redundancy and response diversity (Ostrom, 1995; Low et al, 2003; Duit, 2012). However, where species redundancy is widely recognised and welcomed as contributing to ecological resilience, institutional redundancy may be resisted on the grounds that it leads to fragmentation, duplication, inconsistencies and inefficiencies (Low et al., 2003; Duit, 2012), overcoming perceptions of these problems associated with institutional diversity being recognised as an important challenge (Imperial, 1999).

Young (2010) also recognises the importance of interconnections between diverse institutions in governance frameworks for social systems in conferring resilience to the ecological systems with which they are coupled, these interconnections being analogous to trophic interactions between diverse species in ecosystems. There would thus seem to be a parallel between the recognition that having a high diversity of species, with complex trophic interactions between them, confers ecosystem resilience and that having a high diversity of institutions, with complex interactions between them, confers social system resilience, though the benefits of institutional redundancy are considered by some to be outweighed by the problems it can cause.

It is also important to recognise that ecological resilience is focused not only on the diversity of species, but also on the diversity of functional groups that different species with different roles in the ecosystem constitute. Different categories of institutions with different functions could be considered to be analogous to different functional groups of species, raising the question of whether neo-institutional analysts recognise the importance of diversity amongst different categories of institution, e.g. legal (state), economic (market) and participative (people and civil society). Recognition of the importance of institutional diversity to confer social resilience may not extend to legal institutions, which are, as is discussed earlier in this chapter, considered by many neo-institutionalists to undermine the potential for self-governance in semi-autonomous 'places' unless they passively reinforce decisions, taken on a participative basis by local actors, by providing sanctions. However, the reality is that legal institutions are often associated with obligations that are imposed on local actors, rather than just passively reinforcing local decisions. This is particularly the case with MPAs, as the legal institutions related to them generally include wider-scale strategic conservation objectives and related obligations, as was discussed previously in this chapter in terms of governance challenges related to divergent objectives. This could be one source of resistance to legal institutions amongst neo-institutional analysts of social-ecological systems.

Another source of resistance is related to the view that legal institutions may be too rigid for dynamic social-ecological systems, i.e. whilst laws may promote stability and robustness, their rigidity inhibits the ability of the governance system to quickly respond and adapt to potentially perturbing forces, thereby reducing resilience, though some argue that laws are not necessarily rigid institutions that undermine adaptability and resilience (Ebbesson, 2010; Ruhl, 2012). Young (2010) highlights that striking a balance between rigidity and flexibility is an important challenge to address in the design of institutions that provide for resilience and adaptation. Indeed, it could be argued that legal incentives represent relatively slow institutional variables that can contribute to the resilience of social systems, in the same manner that relatively slow ecological variables can contribute to the resilience of ecosystems (Walker et al., 2006), therefore having some relatively rigid legal incentives can be beneficial, provided a balance is struck between such incentives being slow and rigid enough to withstand the perturbing effects of driving forces, but potentially flexible enough to provide for adaptability. There is common misperception that laws are somehow 'laid down in stone' in that they can take a very long time to revise, but it is important to recognise that laws can be revised in anything from a few days to a few years, depending on the nature of the priority and of the legal system, and that laws are arguably no more inflexible than many other governance institutions.

With regards to striving to achieve the balance discussed above, it is ironic that many neo-institutional governance analysts eschew stability, conferred by 'rigid' laws (Ruhl, 2012) as inhibiting adaptability and thereby resilience, whilst ecologists embrace stability, conferred by complex webs of trophic interaction (McCann et al., 1998; Polis, 1998) amongst a high diversity of species, including slow variables, as promoting adaptability and resilience through functional redundancy and response diversity. Recognition of the potential role of reflexive law is emerging, i.e. legal systems that impose goals and threshold standards, but focus mainly on procedural requirements for deliberative and institutional learning approaches for achieving such goals and standards (Garmestani and Benson, 2013), including 'rolling regulation', which is similar to Dryzek's 'rolling rule regime' discussed above. However, many neo-institutional governance analysts remain resistant to the role of legal institutions in their focus on self-governance, unless they simply reinforce local decisions and do not impose objectives and obligations.

In summary, with regards to the question of whether neo-institutional governance analysts recognise that institutional diversity confers resilience to governance frameworks for social systems, it is clear that they do, including the importance of interconnections between diverse institutions, but that there are reservations that the benefits of institutional redundancy in promoting response diversity may be outweighed by the problems of fragmentation, duplication, inconsistencies and inefficiencies. With regards to the question of whether neo-institutional analysts recognise the importance of diversity amongst different categories of institution, it would seem that there is resistance to the role of legal institutions on the grounds that they are too inflexible and thereby undermine adaptability, and that they are often a vehicle for the authority of the state to undermine self-governance by local actors.

The emphasis of neo-institutional governance analysts remains on self-controlling social systems and resistance to hierarchical approaches which legally impose certain requirements on the outcomes of deliberative processes. This is largely attributable to their focus on self-governance, arguably itself attributable to the alternative participatory development roots of neo-institutionalism. It may also be attributable to the tendency to consider that social systems in a given place should be self-controlling, based on the assumption that ecological systems in a given place are self-controlling, which is itself debatable given the scale and connectivity of marine ecosystems, as discussed in Chapter 3. The focus of social-ecological systems thinking is very much on semi-autonomous or self-organised systems, scale-challenges being addressed by recognition that people and institutions in a given place are nested or embedded in wider-scale institutions, coupled with a faith that facilitated deliberations will lead to consensus and cooperation amongst local people in a given place that yields sustainable outcomes, and that negotiations through horizontal and vertical linkages will address scale challenges. It is argued here that the analogy of self-organising or self-regulating systems may not apply to marine ecosystems, given the strong ecological interconnections related to scale and connectivity, but it certainly does not apply to social systems, given the human interconnections and the related driving forces that MPAs are subject to. This relates to but goes further than arguments that social-ecological systems theorists should not uncritically transfer analogies of rationality and self-regulation from ecological to human systems, as the latter are not necessarily rational or self-regulating (Cannon and Müller-Mahn, 2010).

The emphasis of the co-evolutionary hierarchical governance concept, by contrast, is that whilst some elements of compliance on outcomes may be negotiable through 'rolling rule regimes', certain outcomes required to achieve strategic societal conservation objectives, including those related to equity, are not negotiable. Against this background and building on the co-evolutionary hierarchical governance concept, a key theme that will be pursued later in this book, through the case study findings, is that employing a diversity of institutions from different categories, that combines state, market and civil society approaches to governance, increases both the resilience of the governance framework and the effectiveness of the governed MPA in preventing or minimising over-exploitation, i.e. the key to social resilience is a diversity of institutions representing a combination of all governance approaches, including legal institutions. This can, in turn, lead to the conservation or restoration of the density of populations and diversity of species, and thereby the maintenance or restoration of ecological resilience. This represents a potentially important cause–effect linkage between resilience in social and ecological systems, in that institutional diversity increases not only the resilience of a given social system, but also leads to the increased effectiveness of a given MPA, which, in turn, increases the species diversity and thereby the resilience of the ecosystem.

Chapter 5

Empirical framework for analysing MPA governance approaches

Introduction

The previous chapter focused on a critical analysis of neo-institutional theories on natural resource governance and outlined the co-evolutionary hierarchical governance concept. It is co-evolutionary in that it recognises that top-down, bottom-up and market approaches to governance are interlinked and likely to affect each other's development, e.g. through the influence of local people on higher level institutions. It is also co-evolutionary in that it recognises that increased institutional diversity, increased MPA effectiveness, increased species diversity and increased social-ecological resilience are interlinked, i.e. effective governance approaches have the potential to provide for the co-evolution of both social and ecological systems by promoting resilience through increases in the diversity of institutions and thereby of species. The main focus of this chapter will be to describe the empirical framework that was developed and applied to explore the appropriateness and applicability of the co-evolutionary hierarchical governance concept, through the 20 case studies. This will also enable an exploration of the proposed link between increased institutional diversity, increased resilience of the MPA governance framework, increased MPA effectiveness, increased species diversity and increased resilience of the protected marine ecosystem.

Given the key point of departure between neo-institutional concepts related to place-based self-governance and the co-evolutionary hierarchical governance concept, it clearly would not be appropriate to employ neo-institutional empirical frameworks, such as that for analysing the sustainability of social-ecological systems (Ostrom, 2009) or the fulfilment of related critical enabling conditions (Agrawal, 2001), as these are premised on assumptions that are critically discussed in Chapter 4. However, whilst this book does not share the assumptions underlying these frameworks, it does share recognition of the need for systematic analyses of governance approaches and related issues in different case studies. This chapter will describe the empirical framework that was developed in order to enable such systematic analyses, which draws on and is consistent with the co-evolutionary hierarchical governance concept, in that it explicitly and constructively incorporates the role of the state. This framework was designed to analyse MPA case studies in a manner that is based less on theories and underpinning ideals, and more on the realities of such case studies, including recognition that top-down, bottom-up and market approaches have a combined and inter-linked role to play in addressing the challenges of effectively and equitably governing MPAs.

The roots of this framework may lie in my initial training in biology, which engendered a commitment to empirical research. This includes adopting a systematic approach to case study analyses and applying this approach to a wide range of case studies that are as randomly selected as feasible. Whilst neo-institutionalists also have a commitment to a systematic approach, their case studies tend to be purposefully selected in order to represent contexts where the emphasis is on self-governance by self-organised local actors. This provides for governance analyses to focus on interactions amongst such actors in order to identify combinations of enabling conditions that support the evolution of institutions for sustainable self-governance (Agrawal, 2001).

Neo-institutional case studies also tend to be focused on contexts where sustainable natural resource exploitation is the main objective, rather than wider biodiversity conservation, and tend to exclude, given the focus on self-governance, case studies that are subject to conservation obligations that are, to a degree, imposed by higher levels in the institutional hierarchy. Furthermore, many case studies employing neo-institutional approaches are selected in LEDCs, this arguably being attributable to the focus of such studies on empowering local people, in order to promote locally appropriate and participative development, as discussed in Chapter 4. However, it also means that the findings of such case studies are arguably confined to contexts in LEDCs where self-governance is feasible and locally appropriate, and participative development through sustainable use is the aim. Given that this research project was designed from the outset to analyse MPA case studies which are focused on achieving strategic conservation objectives, employing different governance approaches in a representative range of different national contexts, including MEDCs, clearly a different but nonetheless systematic empirical framework was needed.

Whilst these discussions critique some elements of neo-institutional empirical approaches to analysing the governance of CPRs, a great deal has been learnt from such approaches. In essence, these approaches are focused on analysing how various challenges can be overcome, through interactions amongst local people, in order to develop governance approaches that can provide for the sustainable development of CPRs. These challenges are referred to as collective action problems (CAPs), which are the hurdles that need to be overcome if the potential for commitment, cooperation and compliance is to be developed amongst CPR users and regulators, who are referred to as actors. The tendency of some actors to lack commitment to the sustainable development of CPRs and to behave accordingly in an opportunistic and individualistic manner is generally referred to as free-riding. The social processes that may bind actors in order to address such CAPs and minimise free-riding are often described in terms of social capital. This is a measure of the degree to which actors reach and implement decisions together through their professional and social networks, placing trust in one other, and having confidence that their cooperation will be reciprocated. The social capital approach seeks to modify institutional design and policy processes so that incentive structures are developed that encourage and support actors in overcoming CAPs. Incentive structures are particular types of institution, including shared norms and enforcement laws, which promote commitment, cooperation and compliance amongst resource users (Jones and Burgess, 2005 drawing on Ostrom, 1990; 1998; 1999). Many case studies have been conducted in keeping with this analytical approach, Agrawal (2001) drawing on these in compiling a list of 33 critical conditions that promote the evolution of incentive structures for the sustainable governance of CPRs.

The empirical framework described in this chapter was inspired by this approach, in that it is aimed at providing a systematic approach to analysing case studies. A list of 36 incentives has been developed as the central element of this framework, including incentives that draw directly on neo-institutional analyses. However, given the key point of departure discussed in Chapter 4 between the rationales underlying the bottom-up polycentric and the co-evolutionary hierarchical governance concepts, it was clear that a different empirical framework would be required. Several of the critiques in Chapter 4 were based on the lack of applicability of some of the critical enabling conditions to the governance of human uses of marine ecosystems. More fundamentally, the co-evolutionary hierarchical governance concept does not share the assumptions related to place-based self-governance on which bottom-up polycentric governance concepts are premised. Instead, an empirical framework has been developed that is not constrained by such assumptions, aiming instead to capture the realities of governing MPAs in a variety of representative contexts and that include some form and degree of top-down control in order to achieve strategic wider-scale conservation objectives. This is consistent with calls for research that takes a more 'realist' approach to natural resource governance (e.g. Raik et al., 2008), rather than one that is constrained by ideals as to which governance approach is 'right' or 'best'.

As well as having a theoretical background in such critical analyses of neo-institutional concepts and empirical approaches, the research on which this book is based also has an applied background, building on the current IUCN guidance on the management of MPAs, as was discussed in Chapter 4. This followed discussions with the IUCN in late 2007, which led to a proposal for a study which would systematically analyse and compare MPA governance in a representative range of case studies from around the world, with a view to identifying 'good practice' and assessing the transferability of such governance approaches to other MPAs in comparable contexts. This convergence of a constructive applied perspective and a critical theoretical perspective, with the need for a systematic, widely applicable approach being common to both, led to the development of the empirical framework that is described in this chapter.

A preliminary version of the framework was developed over 2008 and circulated amongst members of the IUCN WCPA-Marine and Global Marine Programme networks for their feedback and suggestions for case studies. The framework was then presented at a workshop on the MPA governance (MPAG) research project, held at the second International Marine Protected Area Congress (IMPAC2, Washington DC, May 2009). This framework was subsequently refined, both in the light of discussions at the workshop and feedback received from other MPA researchers through IUCN and other networks. During 2009, project participants utilised the framework to analyse governance approaches in their case study MPAs, and in October 2009 a case study workshop was held in Lošinj, Croatia, hosted by the Blue World Institute and funded by the United Nations Environment Programme (UNEP). This workshop focused on 20 case study sessions based on the application of the empirical framework, and a resulting UNEP technical report based on these analyses and discussions was subsequently published (Jones et al., 2011).

This report involved a combination of analyses undertaken by the case study representatives before the workshop and further analyses undertaken during the workshop, through collaborative discussions of each case study, and after the workshop by the MPA governance analysis researchers: Peter Jones (author of this book), Wanfei 'Feifei' Qiu and Elizabeth De Santo. The empirical framework employed in this book has been refined and revised in the light of learnings from its application to the 20 case studies on which this book draws. In particular, the list of 33 incentives used in the analysis of these 20 case studies has been revised to a list of 36 incentives, along with other minor changes to the empirical framework, such as the addition of some metrics. This chapter will describe the refined and revised empirical framework that will be used in future case studies whilst the next chapter will draw on the findings of the 20 case studies, as these are not significantly affected by the revisions to the framework. The framework and discussions of the findings build on Jones et al. (2013a; 2013b).

Positionality

Before going on to describe the empirical framework, it is important to consider how my world view, background and priorities could be considered to have influenced this research, and to make any related assumptions underlying this research clear. I have always had a fascination with the sea, having grown up in the south coast town of Poole, England, which has a very large estuarine harbour as well as open sandy beaches and chalk cliff coastlines. My youth included many maritime activities, ranging from angling, playing and relaxing on the beach, body boarding, snorkelling and sailing, through to the door-to-door selling of crabs bought at the quayside from local potting vessels and 'fishing boat watching', i.e. a maritime version of trainspotting involving collecting the numbers and names of local fishing boats. I went on to gain a degree in biological sciences, specialising in marine ecology, and a masters that combined studies of applied marine ecology and marine policy. My PhD focused on the value conflicts underlying the objectives, selection and management of MPAs, with a particular focus on the UK, involving inter-disciplinary research drawing on many different theoretical and applied literatures, from environmental ethics, through to marine ecology and marine policy.

I can identify with both conservationist and utilitarian views of the sea, my personal and academic interest in MPAs being driven mainly by the former but with a strong appreciation of and respect for the latter. I am inclined towards the marine ecologists' views on the divergent perspectives discussed in Chapter 3, but I also completely appreciate that others are mainly driven by utilitarian values, recognising that we all hold a spectrum of values, with different priorities coming to the fore in different contexts. I am also a humanitarian and an anthropocentrist, i.e. the welfare and happiness of humans is my main focus of moral concern, but this is mirrored by the view that the well-being of the human race is inextricably intertwined with the health of ecosystems that are our life support systems, i.e. ecologically enlightened anthropocentrism.

In the previous chapter, it was considered, in relation to Foucalt's work (Dreyfus and Rabinow, 1982), that many of the references cited in support of MPA objectives in Chapter 2 could be argued to represent normative science, i.e. based on subjective conservationist values rather than solely on objective science. Hilborn (2006) criticises such papers as representing a belief in and advocacy of no-take MPAs based on faith, rather than representing an objective case for such designations based on properly tested and critically reviewed scientific evidence. Others, however, have long argued that conservation biology is an inherently normative science of which advocacy for the preservation of biodiversity is a necessary and important part, particularly given the pervasiveness of uncertainty (Barry and Oelschlaeger, 1996). These views are consistent with recognition of embedded normativities discussed in Chapter 3. The role of scientists as advocates of no-take MPAs is an important issue in this respect, given arguments that the distinction between science and advocacy is becoming increasingly blurred but also that scientists have a moral obligation to inform society if they believe that the earth's ecological life support systems are being damaged (Jones, 2006).

Whilst I agree with the view that the papers cited in Chapter 2 could be considered as being normative in their support for no-take MPAs, I also consider that: (a) declines in the health of marine ecosystems represent *real* rather than constructed concerns; (b) biologists investigating such declines cannot base their arguments solely on objective, value-free, positivist science; and (c) MPAs are a necessary element of a precautionary approach to addressing concerns about such declines, given the high importance of marine ecosystems coupled with the high degrees of uncertainty related to our knowledge of them. Considering this view, I could reasonably be described as an advocate for MPAs, including no-take areas, be these entire designations or zones within designations, as I consider that no-take areas are a necessary measure, amongst others, to restore the health of marine ecosystems and address related scientific and societal concerns, and I think that there is sufficient objective scientific evidence to support this view, accepting the need for precaution in the face of high stakes and high uncertainty,[6] though I also recognise the importance of addressing equity issues raised by MPA initiatives.

Based on this position, the empirical framework described in this chapter is explicitly 'directional', in that it is premised on an acceptance of the need for effective MPAs to address growing societal concerns and related uncertainties, and is specifically aligned towards the fulfilment of MPA objectives, particularly those discussed in Chapter 2. From a governance analysis perspective, by contrast, the MPAG empirical framework is not premised on an acceptance that any particular governance approach is 'right' or 'best', as was discussed in Chapter 4.

The 20 MPA case studies on which this book is focused have been analysed from the perspectives of the case study contributors, who come from a variety of backgrounds and are affiliated to MPA management agencies, as well as NGOs and academic institutes who are involved in the MPAs. Whilst the analyses are from the perspective of these case study contributors, rather than being based on a full ethnographic study of the perspectives of different actors, each contributor has extensive knowledge of their MPA from practitioner, research and/or NGO perspectives, though it is important to recognise that the case study analyses will, to varying degree, reflect the positionality of the case study contributors and of the researchers on the MPA governance project who have also contributed to these analyses. However, no study is free of positionality issues and the consistent use of the MPAG empirical framework, which is not constrained by certain theories or their underlying ideals, should offset some positionality issues.

The empirical framework

The basic elements of the empirical framework are outlined in Box 1 and will be described in detail in this section.

Box 1 Outline of the MPAG empirical framework

- Context, including metrics:
    - Name of MPA
    - Area of MPA
    - Coastline length (where applicable)
    - Per capita Gross Domestic Product (GDP)
    - GDP growth rate
    - Main economic sectors (relative employment and GDP contributions)
    - Unemployment rate
    - State capacity
    - Human Development Index (HDI)
    - Population below poverty line
- Objectives of the MPA
- Drivers and conflicts
- Governance framework/approach
- Effectiveness (0–5)

- Incentives employed and needed:
  - Economic
  - Interpretative
  - Knowledge
  - Legal
  - Participative
- How the incentives are combined, relative importance, etc.
- Cross cutting themes:
  - Role of leadership
  - Role of non-governmental organisations
  - Equity issues

Context

When analysing the effectiveness of different approaches to governing MPAs, it is important to consider the local and national context, as this can be a significant determinant of why particular governance approaches are appropriate and effective, i.e. 'context counts', though it has been neglected by many analysts of CPR governance regimes (Edwards and Steins, 1999). The local and national institutions in which a given MPA is embedded (McCay, 2002) represent key elements of the context. The main aim of this section is to provide a brief overview of the MPA in relation to the socio-economic and political situation at a national level, along with any particularly significant contextual elements at a local level. The importance of the first three basic descriptive headings is reasonably self-evident, but it is worth briefly discussing the role of the metrics employed in the context section.

Per capita gross domestic product (GDP), along with the GDP growth rate, is arguably the key metric that most modern governments have a particular focus on. The former gives an indication of the economic wealth of people in that country, which can have a direct bearing on the direct and indirect uses of marine areas that are likely to be their priority. The latter gives an indication of how fast the economy is expanding or contracting, this often also being a key factor that affects the direct and indirect uses that actors may prioritise. The relative contribution that different economic sectors make to GDP and/or employment is a good indication of the relative importance of agriculture (including fisheries), the service sector (private sector that does not produce goods, e.g. finance, tourism, etc.) and industry. The unemployment rate is also a useful indicator of the state of the economy and the priority that may be attached to job creation.

The percentage of the population below the poverty line is a relative measure, depending on the wealth of the country and the cost of living. It is a useful indicator of the proportion of the population that is economically very disadvantaged, for whom livelihood and economic development opportunities are a very high priority. All of these metrics are usually readily available at a national level from online sources such as the CIA World Factbook. They may not be available for the local area of the MPA, but it is usually feasible to give some indication of the state and nature of the local economy relative to the national economy, e.g. poorer/richer, lower/higher unemployment, importance of particular sectors.

State capacity is the average of the indicators for six dimensions of governance[14] developed and analysed by the World Bank: voice and accountability; political stability and absence of violence; government effectiveness; regulatory quality; rule of law; and control of corruption (Kaufmann et al., 2009). The average of these six indicators serves as an overall indicator of the national capacity of the state for enabling stable governance that could contribute to the effective achievement of strategic societal objectives, such as those for MPAs. The Human Development Index (HDI) is a combination of metrics that indicate societal 'health', particularly education, life expectancy and standard of living, these being important social contextual factors for MPAs. The closer to 1, the more developed a country is in these respects, the ranking indicating their relative position in the HDI of the entire world's ranked countries.[15]

Any one of these metrics, alone, could be criticised on the grounds that it only gives a snapshot of a complex context from a single perspective. Each of these metrics are also subject to criticisms on the grounds of their rationale and/or the partial and reductive way in which they are calculated, reducing socially, economically, politically and culturally complex issues to a single number. They are, however, generally readily available and provide a rapid and straightforward means of enabling the analyses of MPA governance approaches to be considered in the national and local context, recognising that local metrics are generally less readily available. Taken as a group, they also go some way to offsetting the criticisms of any one metric on the grounds that it is too 'blinkered'. Whilst it is important to bear the limitations and criticisms of such metrics in mind, it is also important to recognise that each of them is the result of massive data gathering and analysis efforts, any one of which would be way beyond the remit and expertise of scientists studying MPA governance issues. As such, it is argued that this group of metrics can yield some very interesting insights for relatively little effort, allowing the context to be taken into account when analysing different MPA governance approaches. They enable the MPA governance approach to be considered in the wider national governance context and relative to other case studies in other countries, recognising the importance of context in such governance analyses (Jones and Burgess, 2005).

Objectives of the MPA

All MPAs should have clear and specific conservation objectives that the 'protected' status is aimed at achieving. These generally fall into one or more of the previously discussed nine categories (Chapter 2) and tend to be more focused on conservationist values, where users of MPAs tend to be more focused on utilitarian values (as was discussed in Chapters 3 and 4), potentially leading to basic conflicts, which represent important governance challenges. Most MPAs have specific objectives that form an important basis of governance analyses. Participants were requested to specify the objectives of the MPA, be these formally stated in a policy and/or legal document, or informally outlined but still recognised.

Drivers and conflicts

The objectives of a given MPA are often related to strategic objectives at a national, regional and international level, and these may be subject to related policy priorities and legal obligations that can have a significant influence on governance processes. The fulfilment of such conservation objectives, priorities and obligations generally has to address pressures and impacts from various activities, such as tourism and fishing, and it is important to establish the main sectoral activities that may need to be subject to some form of steer, restrictions and controls in order to achieve the MPA's objectives. It is widely acknowledged that protected areas in general are being increasingly influenced by the global forces of economic development and socio-political change (Dearden et al., 2005; Büscher and Whande, 2007), these also being discussed in terms of the root causes of biodiversity loss (Wood et al., 2001). Such forces are increasingly recognised as the sorts of human interconnections that pose scale challenges for protected area governance, as was discussed in Chapter 4.

There are particular driving forces that MPAs must face, such as increasing human populations, both from local population growth and inward migration, increasing demands from globalised fish and tourism markets, and the increasing aspirations of people to improve their living standards beyond subsistence livelihoods. Such driving forces could be considered in terms of disincentives that could potentially undermine the fulfilment of conservation objectives, but for the purposes of this analysis, such disincentives are discussed in terms of the driving forces that incentives designed to support MPAs must address and withstand. Participants were requested to describe the main activities that need regulating to achieve the MPA's objectives, the trends in these activities and the forces that are driving these trends. These driving forces are discussed further in relation to their impacts and influences on the MPA case studies in Chapter 6.

Having identified the objectives of a given MPA, the obligations driving these objectives, the main sectoral activities that could impact MPA features and potentially undermine the fulfilment of these objectives, and the forces driving these activities, this logically leads to an initial assessment of the main conflicts that MPA governance needs to address. Conflicts occur when the sensitive habitat and species features of a MPA are exposed to an activity that can impact their condition, thereby undermining the fulfilment of the MPA's conservation objectives. MPAs aim to proactively or reactively control and restrict such activities, often including addressing the driving forces behind such activities, in order to reduce or control such impacts and thereby ameliorate or avoid the conflicts. Participants were requested to identify the main conflicts, particularly the activities that need restricting to promote the achievement of the MPA's objectives and fulfilment of any related objectives, including the role of the driving forces behind such activities.

Governance framework/approach

MPAs are often linked to some form of institutional hierarchy, in that they are recognised as an element of a legal and/or policy[16] framework. These are often hierarchical, in that the specific legal instrument or policy document by which the MPA is officially designated is legally/officially recognised as contributing to a national or state law/policy, which is itself legally/officially recognised as contributing to a federal, regional and/or international convention/policy. Such institutional linkages to a legal/policy hierarchy are often particularly important in relation to the objectives of a given MPA, which often contribute to the fulfilment of national or state obligations, which may themselves fulfil regional, federal and/or international obligations. Participants were requested to describe the legal/policy hierarchy of which the MPA is a component, including any related obligations and their hierarchical linkages. A full legal and policy analysis in this respect is neither warranted nor feasible, the essential aim being to gain a basic understanding of the legal and/or policy framework of which the MPA is a component and any related vertical linkages in the hierarchy, particularly obligations and the related potential for legal interventions.

This provides the setting for an account of how the conflicts that arise in a given MPA were addressed through the governance framework and approaches in order to ensure the effectiveness of the MPA. This usually includes a description of the approach taken to control activities that could undermine the fulfilment of the MPA's objectives, the challenges such approaches faced and their success, or otherwise, in addressing such challenges and promoting effectiveness.

Effectiveness

As was previously discussed, the 'bottom line' for these governance analyses is whether the governance approach provides for the achievement of a given MPA's objectives. A key focus of the workshop was to assess effectiveness in terms of the degree to which the local impacts of certain activities which could undermine the achievement of MPA objectives have been reduced or controlled in order to ameliorate or avoid conflicts. This involves an assessment of the proportion of impacts that are being addressed (or not) and the degree to which they are being addressed (or not). This is focused on the effectiveness of the MPA governance incentives in addressing impacts from local and incoming users, and from surrounding and upstream uses, recognising that this has the potential to enhance ecosystem resilience to wide-scale impacts that are beyond the control of individual MPAs, particularly those related to climate change, e.g. increases in seawater temperature leading to coral bleaching. Where such wider-scale impacts are impacting a MPA, this should be highlighted but should not reduce the effectiveness score, as this is focused on those impacts related to local and surrounding uses which can be addressed within the MPA's governance framework. A scale of 0–5 was used to represent effectiveness:

0   No use impacts addressed; MPA designation may even have increased impacts by undermining previous governance institutions
1   Some impacts beginning to be slightly addressed
2   Some impacts partly addressed but some impacts not yet addressed
3   Some impacts completely addressed, some are partly addressed
4   Most impacts addressed but some not completely
5   All impacts from local activities completely addressed.

The assignment of an effectiveness score was based solely on a qualitative judgement through discussions with case study representatives, rather than on a detailed quantitative analysis. Whilst it is recognised that analyses of the effectiveness of MPAs in achieving their objectives needs to take account of the ambitiousness of the objectives themselves (Jones, 2001), this analysis was confined to a focus on the impacts related to the actual objectives of each MPA. For future studies, more account could be taken of the ambitiousness of the objectives, including the proportion of the MPA designated as no-take in order to achieve ecosystem and fisheries restoration objectives, compared to the proportion with only partial protection from some activities. However, for the purposes of this initial study, the focus was on an estimate of effectiveness based on the actual conservation, restoration and sustainable use objectives of the MPA in question.

Five categories of incentives

This is where the theoretical discussions on different governance approaches in Chapter 4 meet the realities of the MPA case studies. These discussions were focused on the co-evolutionary hierarchical governance concept, which represents a combination of governance approaches, i.e. top-down, bottom-up and market. Such approaches can be considered in terms of various institutions, as defined in the previous chapter. This empirical framework focuses on incentives, which are defined as 'particular institutions that are instrumentally designed in relation to a MPA to encourage actors to choose to behave in a manner that provides for certain strategic policy outcomes, particularly biodiversity conservation and restoration objectives, to be achieved'. Five categories of incentives are employed in this empirical framework to analyse MPA governance. These can be considered as contributing to one or more of the three sources of governance steer discussed in Chapter 4 (Table 8), as summarised in Table 11.

The case study participants were asked to identify which of these incentives were currently being used in the governance of the MPA on which they were focused, including a description of the way in which each incentive was being used. This included indications of how different incentives were inter-linked and combined. Participants were also asked to indicate which particular incentives from this list are needed or could be useful to address weaknesses in the current governance framework and thereby improve the effectiveness of the MPA. A total of 36 individual incentives have so far been identified and these are listed in Table 12 and discussed in detail in Chapter 7 in relation to their use in the 20 case studies.

Table 11 Five categories of incentives

Incentive category	Definition (number of incentives in this category employed in this framework)	Relevant governance mode
Economic	Using economic and property rights approaches to promote the fulfilment of MPA objectives (10)	Market-based
Interpretative	Promoting awareness of the conservation features of the MPA, the related objectives for conserving them and the approaches for achieving these objectives, and promoting support for related measures (3)	Supporting all three approaches
Knowledge	Respecting and promoting the use of different sources of knowledge (local-traditional and expert-scientific) to better inform MPA decisions (3)	Supporting all three approaches
Legal	Establishment and enforcement of relevant laws, regulations etc. as a source of 'state steer' to promote compliance with decisions and thereby the achievement of MPA obligations (10)	Top-down (state steer)
Participative	Providing for users, communities and other interest groups to participate in and influence MPA decision-making that may potentially affect them, in order to promote their 'ownership' of the MPA and thereby their potential to cooperate in the implementation of decisions (10)	Bottom-up (people steer)

Table 12 List of incentives in each category

Incentive category	Incentives (total 36)
Economic (10)	Payments for ecosystem services (PES) Assigning property rights Reducing the leakage of benefits Promoting profitable and sustainable fisheries Promoting green marketing Promoting alternative livelihoods Providing compensation Reinvesting MPA income in local communities Ensuring sufficient state funding Provision of NGO and private sector funding
Interpretative (3)	Raising awareness Promoting recognition of benefits Promoting recognition of regulations and restrictions
Knowledge (3)	Promoting collective learning Agreeing approaches for addressing uncertainty Independent advice and arbitration

Table 12 continued

Incentive category	Incentives (total 36)
Legal (10)	Hierarchical obligations Capacity for enforcement Penalties for deterrence Protection from incoming users Attaching conditions to property rights Cross-jurisdictional coordination Clear and consistent legal definitions Clarity concerning jurisdictional limitations Legal adjudication platforms Transparency and fairness
Participative (10)	Rules for participation Establishing collaborative platforms Neutral facilitation Independent arbitration panels Decentralising responsibilities Peer enforcement Building social capital Bracing linkages Building on local customs Potential to influence higher institutional levels

Five categories of governance approach

As well as the five categories of particular governance incentives, five categories of overall MPA governance approach were employed. These approaches partly represent different degrees and types of decentralisation, which is defined as 'the transfer of authority from central government to lower-level government levels, quasi-independent government organisations, NGOs or the private sector'. According to Rondinelli (2000) and Oxhorn (2004), there are different types of decentralisation that allocate varying degrees and forms of autonomy to subnational government levels, quasi-independent government organisations, NGOs or the private sector:

- Deconcentration – the transfer of power for implementing decisions, but not for making decisions
- Delegation – transfer of some decision-making powers with a degree of control from the central government over key aspects of policy
- Devolution – the transfer of maximum feasible, but not necessarily total, decision-making powers.

Decentralisation is an important concept in governance that can be interpreted, from a more radical neoliberal perspective, as a means for the state to relinquish power to NGOs and the private sector in order to remove barriers to economic development. From the neo-institutional perspective discussed in Chapter 4, decentralisation is a fundamentally important means of creating quasi-autonomous decision-making units whose authority is not undermined by interventions from the higher government levels, particularly in the context of place-based self-governance. This analysis will consider decentralisation from the perspective of the co-evolutionary hierarchical governance concept, in keeping with the definition and levels of decentralisation that are discussed above, whereby the state may deconcentrate, delegate or devolve power, but it should not relinquish it by totally transferring power, as scale challenges such as those posed by MPAs require some form and degree of state control in order to effectively and equitably address them. Accordingly, five categories of MPA governance approach are employed in the empirical framework, as summarised in Table 13.

The actual characteristics of MPAs assigned to these categories will be discussed in detail in Chapter 6, as will the related governance issues.

Table 13 Five categories of MPA governance approach

MPA governance approach	Characteristics
Governed primarily by the state under a clear legal framework	Decisions are taken by the state with some deconcentration or delegation of power to lower level government and quasi-independent government organisations, which generally only consult local users and other stakeholders on decisions taken at a higher state level
Governed by the state with significant decentralisation and/or influences from private organisations	Implementation is deconcentrated to lower level government, quasi-independent government and private organisations along with the delegation of some decision-making powers, with central governments maintaining some degree and form of control over implementation and decision-making
Governed primarily by local communities under collective management arrangements	MPAs instigated on a bottom-up basis by local users, often through local organisations, with many implementation and decision-making remaining delegated to local users/organisations, but often requiring some degree of state support for enforcement and therefore also involving some influence by central governments
Governed primarily by the private sector and/ or NGOs who are granted with property rights and associated management rights	MPAs instigated by organisations who may, or may not, represent local users and who are granted with the majority of decision-making powers and implementation responsibilities, but often still requiring some degree of state support for enforcement, though central government influence is generally limited to conditions attached to property and associated management rights, coupled with recourse to withdraw such rights if conditions are not fulfilled
No clearly recognisable effective governance framework in place	Paper MPAs with no effective incentives to promote the achievement of MPA objectives or fulfilment of related obligations, the central state or lower government levels often lacking the political will for effective MPAs and sometimes being involved, in partnership with the private sector, in development proposals that significantly undermine such fulfilment

Cross-cutting themes

Particular governance issues emerge in many MPA case studies and are more widely recognised as key themes that emerge in many natural resource governance debates. For the purposes of this analysis of MPA governance, three such cross-cutting themes will be focused on in the discussions of the MPA case study findings – role of leadership, role of NGOs and equity issues.

Role of leadership

The role of key individuals who have entrepreneurial and innovation skills, and who are respected by and have influence amongst people, is recognised as a key attribute that can play a particularly important role in promoting effective governance in social-ecological systems (Folke et al., 2005; Ostrom, 2009; Ostrom and Cox, 2010). Leaders can play an important role amongst local user communities, amongst NGOs and/or amongst government organisations, the latter being particularly important in promoting the political will to support the development of effective MPA governance frameworks. Leaders can also play an important role in bridging different organisations, i.e. in developing vertical and horizontal linkages (discussed in Chapter 4) and brokering negotiations (Bodin et al., 2006), in bringing key knowledge to governance processes and in promoting collective learning. The emphasis of neo-institutionalists is on the important role of leadership in promoting self-governance, but it is just as important in the context of the co-evolutionary hierarchical governance concept.

Role of NGOs

NGOs are recognised as playing an increasingly important role in governance, particularly given the growing focus on the role of civil society discussed in Chapter 4. Civil society can be considered as the space between individual people and the state, and NGOs play a key role in organising this space, building groups, coalitions, networks and bridges. In this respect, NGOs may assume the leadership and linkage development roles discussed above in relation to the role of leadership, though their roles are not confined to these. They may also work alongside state, private, academic or local community actors, to support MPA governance frameworks in order to provide for the better achievement of MPA objectives. They are also often responsible for galvanising both public opinion and political will towards establishing MPAs, for example with the continued push towards networks of MPAs and large no-take MPAs. Both private and academic bodies, e.g. dive tourism operators and marine research institutes, can fulfil roles normally associated with NGOs, such as advocating MPAs and supporting their designation and management.

NGOs can also have an important role in promoting the interests of local users, particularly in countries that have a relatively weak history of and capacity for public participation (Mol and Sonnenfeld, 2000). This role, however, is mirrored by concerns, such as those discussed in Chapter 2 in relation to the MPA objective of controlling tourism, that big international NGOs (BINGOs) can be a part of the 'unholy alliance of global-level environmental and commercial interests' (Homewood et al., 2009, 25) that can lead to the political marginalisation and geographic displacement of indigenous communities (Chapin, 2004; Adams and Hutton, 2007; Brockington et al., 2008, 146–147), thereby raising social justice issues, such as those discussed below. In a related sense, NGOs may play a pivotal role in 'public–private partnerships' related to MPAs, whereby they may broker decentralisation arrangements through property and management rights agreements, sometimes themselves taking on such rights and related roles. However, such approaches could be criticised on the grounds that NGOs may assume a quasi-state role, whilst having fewer requirements for accountability than the state. These concerns are consistent with critiques of participative development, previously discussed in terms of the risks of parochialism (Chapter 4), i.e. (a) that NGOs tend to homogenise as 'community' what are actually heterogeneous 'communities' (Cleaver, 2001); (b) that NGOs may represent key actors in the 'new tyranny' of participative development (Cooke and Kothari, 2001), including allowing elite capture of participative processes (Platteau, 2004); (c) that the effectiveness of NGOs in promoting participation and alleviating poverty lacks empirical evidence (Platteau, 2004); (d) that NGOs are focused on imposing their own values rather than community values; and (e) that NGOs tend to lack legitimacy and accountability (Ebrahim, 2003).

Such concerns aside for now, NGOs are recognised as playing various important roles in the governance of social-ecological systems (Folke et al., 2005; Ostrom, 2009) and protected areas (Dearden et al., 2005), including MPAs (e.g. Johannes, 2002; Crabbe et al., 2010; De Santo, 2012). Such roles include campaigning for MPAs, raising awareness, facilitating deliberations, providing training, providing or facilitating access to funding, promoting knowledge exchanges, brokering decentralisation arrangements in co-management and acquiring marine areas for management as owners. As with leadership, the emphasis of neo-institutionalists is often on the role of NGOs in promoting participation and self-governance, but NGOs can also play important roles in governance from the perspective of the co-evolutionary hierarchical concept. Overall, it is important to recognise that NGOs can have a variety of roles, aligned with top-down, bottom-up or market-focused approaches, and that these roles can be very significant in the governance of MPAs.

Equity issues

Ensuring the equitable sharing of costs and benefits associated with protected areas is becoming an increasingly important issue (Adams, 2004; Blaustein, 2007). There is growing recognition in the post-'fortress conservation' co-management paradigm that protected areas can raise major human rights and social justice concerns, amidst related concerns that poor, disadvantaged and marginalised people can also represent a major threat to the fulfilment of protected area conservation objectives, as they struggle to survive and improve their standard of living, often involving potentially incompatible uses. There are also concerns, as discussed in Chapter 4, that co-management and community-based governance approaches can lead to the reinforcement of inequities and the increased marginalisation of disadvantaged groups, as locally powerful groups and elite individuals can dominate local decision-making processes.

The governance of MPAs is recognised as potentially raising important equity issues, though these are often neglected in MPA analyses due to the dominance of biology, ecology and economics in this field (Blount and Pitchon, 2007). Equity issues are, however, central to the reactions of MPA users, such as fishermen, to MPA proposals and designations and therefore research on such issues should be a priority (Jones, 2009). Blount and Pitchon (2007) report that fishermen are often more committed to fishing as a 'way of life' rather than on the basis of economic rationalism, and therefore often view equity issues in terms of the effects of MPAs on their way of life rather than purely in terms of the distribution of costs and benefits. Recognition that fishing is often more than just a means of subsisting and earning a living is therefore particularly important in research on MPA equity issues, as alternative livelihoods and economic compensation may not fully address the loss of fishing as a traditional way of life.

The challenge is that MPAs often involve restrictions on fishing, particularly no-take MPAs, therefore basic conflicts and related equity issues are often bound to be raised. However, the need to address social justice issues, whereby certain users, such as fishermen, inequitably bear many of the costs of MPA restrictions whilst gaining few of the benefits, have to be balanced against the need to address environmental justice issues, whereby wider direct and indirect users of a given marine area have to bear the environmental costs associated with the impacts of fishing (Jones, 2009). Achieving a balance between social and environmental justice issues often involves trade-offs between equity and effectiveness (Halpern et al., 2013). Furthermore, what may be argued to represent 'traditional' fishing may actually represent or could develop, in the face of driving forces, into relatively intensive modern fisheries for export markets, as was discussed in Chapter 2 in relation to the MPA objective of maintaining traditional fisheries. Whilst addressing inequities and related justice issues can help address basic conflicts, given that equity and effectiveness are inter-related, balancing the need to provide for effectiveness and minimise inequities is often extremely challenging in MPA governance.

Summary

This chapter has described the empirical framework through which the issues introduced in the previous four chapters will be explored in the 20 case studies. This empirical framework is aimed at assessing the effectiveness of MPAs in addressing the growing societal and scientific concerns discussed in Chapter 1 by fulfilling the objectives discussed in Chapter 2. Recognising the differences and divergences discussed in Chapter 3, the application of the empirical framework described in this chapter to the 20 case studies will enable analyses of how different incentives contribute to effectiveness in different MPA governance approaches, drawing on the co-evolutionary hierarchical governance concept discussed in Chapter 4.

The following chapter will overview some general findings from and patterns amongst the 20 case studies, whilst Chapter 7 will describe how each of the 36 incentives has been applied in different case studies. Chapter 8 will then discuss the cross-cutting themes in relation to the case study findings and the importance of combining incentives in relation to the concept of social-ecological resilience through institutional diversity, leading to some overall conclusions.

Chapter 6

Overview of case studies

Introduction

The previous chapter described the process by which the case studies were sought and selected, introduced the five categories of incentives (Table 11), the 36 incentives under these categories (Table 12) and the five governance approaches (Table 13), and outlined the types of driving forces that MPA governance generally have to address. This chapter will begin by introducing the 20 MPA case studies, followed by an overview of the main patterns that emerge from the case studies in different governance approach categories, and ending with a brief discussion of the main driving forces that have had to be addressed by the governance framework for the case study MPAs. The case studies are listed in Table 14, with further details in Table 15, and their global distribution is illustrated in Figure 7.

It is not feasible to give a full account of each case study, for which readers are referred to the case study 'essence' reports in Appendix 1 of Jones et al. (2011) and to the paper for each case study, referenced in Table 14, which gives a fuller account based on an analysis employing the MPAG framework. Relevant details of the various case studies will be specifically drawn on to illustrate the subsequent discussions on various governance issues in this and later chapters. The case studies are listed in Table 15 under the five governance approach categories, including some of the contextual metrics introduced in Chapter 5. In the interests of brevity, the MPA case studies will be referred to by their short name, the reference for each being listed in Table 14. Where case study findings are reported from sources other than these MPAG references, these other sources will be specifically cited. Some insights were also gleaned from discussions at the case study sessions in the MPAG workshop (see Chapter 5) and my own knowledge of these case study MPAs, and if no other source is apparent, i.e. the information is not in the source listed for that case study (Table 14) and not cited to another source, the views are solely mine.

Table 14 List of MPA case studies including main contacts and key reference

Case Study MPA name	Country	Short name	Main case study contacts	Reference
Great Barrier Reef Marine Park	Australia	GBRMP	Jon Day and Kirstin Dobbs, Great Barrier Reef Marine Park Authority	Day and Dobbs (2013)
Darwin Mounds Marine Special Area of Conservation	UK	Darwin Mounds	Elizabeth De Santo, Marine Affairs Program, Dalhousie University	De Santo (2013b)
North East Kent European Marine Site	UK	NE Kent	Tom Roberts, Durrell Institute of Conservation and Ecology, University of Kent	Roberts and Jones (2013)
Wash & North Norfolk Coast European Marine Site	UK	Wash	Peter Jones, University College London and Tom Roberts, Durrell Institute of Conservation and Ecology, University of Kent	Jones (2011)
National Marine Sanctuaries	USA	US NMSs	Elizabeth Moore, National Oceanic and Atmospheric Administration (NOOA)	Moore (2011)
California MPAs under the Marine Life Protection Act	USA	California	Emily Saarman, Marine Life Protection Act Initiative, California Natural Resources Agency and Mark Carr, University of California Santa Cruz	Saarman and Carr (2013)
Sanya Coral Reef National Marine Nature Reserve	China	Sanya	Wanfei Qiu, University College London	Qiu (2013)
Seaflower MPA	San Andres Archipelago, Colombia	Seaflower	Elizabeth Taylor, CORALINA and Marion Howard, Brandeis University	Taylor et al. (2013)
Galápagos Marine Reserve	Ecuador	Galápagos	Veronica Toral-Granda, independent consultant, Gonzalo Banda-Cruz and Scott Henderson, Conservation International	Jones (2013b)
Karimunjawa Marine National Park	Indonesia (Coral Triangle)	Karimunjawa	Stuart Campbell, Wildlife Conservation Society	Campbell et al. (2013)

Table 14 continued

Case Study MPA name	Country	Short name	Main case study contacts	Reference
Wakatobi National Park	Indonesia (Coral Triangle)	Wakatobi	Julian Clifton, University of Western Australia	Clifton (2013)
Tubbataha Reefs Natural Park	Philippines (Coral Triangle)	Tubbataha	Marivel Dygico, WWF, Angelique Songco, Tubbataha Management Office and Alan White, The Nature Conservancy	Dygico et al. (2013)
Ha Long Bay World Heritage Site	Vietnam	Ha Long Bay	Bui Thi Thu Hien, IUCN Vietnam	Hien (2011)
Os Miñarzos Marine Reserve	Spain	Os Miñarzos	Lucia Perez de Oliveira, University College London	Perez de Oliveira (2013)
Isla Natividad MPA	Mexico	Isla Natividad	Wendy Weisman and Bonnie McCay, Rutgers University	Weisman and McCay (2011)
Great South Bay Marine Conservation Area	USA	Great South Bay	Carl LoBue and Jay Udelhoven, The Nature Conservancy	LoBue and Udelhoven (2013)
Chumbe Island Coral Park	Tanzania	Chumbe	Sibylle Riedmiller, Chumbe Island Coral Park and Lina Mtwana Nordlund, University of Gothenburg	Nordlund et al. (2013)
Baleia Franca Environmental Protection Area	Brazil	Baleia Franca	Hietor Macedo, Chico Mendes Institute of Biodiversity Conservation and Melissa Vivacqua, Federal University of Santa Catarina	Macedo et al. (2013)
Pirajubaé Marine Extractive Reserve	Brazil	Pirajubaé	Leopoldo Gerhardinger, University of Campinas and Renata Inui, Coastal and Marine Studies Association	Gerhardinger et al. (2011a)
Cres-Lošinj Special Zoological Reserve	Croatia	Cres-Lošinj	Peter Mackelworth, Blue World and Draško Holcer, Croatian Natural History Museum	Mackelworth et al. (2013)

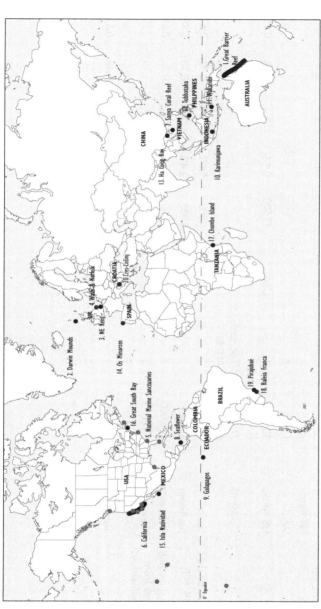

Figure 7 Map of the 20 MPA governance analysis case studies. 1 Great Barrier Reef Marine Park (Australia); 2 Darwin Mounds Marine Special Area of Conservation; 3 North East Kent European Marine Site; 4 Wash & North Norfolk Coast European Marine Site (UK); 5 National Marine Sanctuaries (a network of MPAs with locations shown in grey colour); 6 Californian MPAs under the Marine Life Protection Act (US); 7 Sanya Coral Reef National Marine Nature Reserve (China); 8 Seaflower MPA (Colombia); 9 Galápagos Marine Reserve (Ecuador); 10 Karimunjawa Marine National Park; 11 Wakatobi National Park (Indonesia); 12 Tubbataha Reefs Natural Park (Philippines); 13 Ha Long Bay UNESCO World Heritage Site (Vietnam); 14 Os Miñarzos Marine Reserve of Fisheries Interest (Spain); 15 Isla Natividad MPA (Mexico); 16 Great South Bay Private Marine Conservation Area (US); 17 Chumbe Island Coral Park (Tanzania); 18: Baleia Franca Environmental Protection Area; 19 Pirajubaé Marine Extractive Reserve (Brazil); 20 Cres-Lošinj Special Zoological Reserve (Croatia). Reprinted from Jones et al. (2013a) with permission from Elsevier.

Table 15 Governance approach categories, MPA case studies and some key metrics

MPA governance approach	Short name	Area (km²)	Effectiveness*	Country	National Per Capita GDPUS$*	GDP annual growth rate*	Pop. below poverty line*	State capacity*	Human development index (world ranking)*
(I) Managed primarily by the government under clear legal framework	GBRMP	344,400	3	Australia	38,200	2.4%	13%	1.65	0.935 (2)
	Darwin Mounds	1,380	3	UK	36,700	0.7%	14%	1.48	0.847 (26)
	NE Kent	23	3						
	Wash	1,078	3						
	US NMSs	407,740 (14 MPAs)	3	USA	47,500	0.4%	15%	1.36	0.899 (4)
	California	21,927 (124 MPAs)	Too early to assess						
(II) Managed by the government with significant decentralisation and/ or influences from private organisations	Sanya	56	2	China	6,000	9.0%	13%	-0.47	0.655 (89)
	Seaflower	65,000	1	Colombia	9,200	2.4%	34%	-0.38	0.685 (79)
	Galápagos	138,000	1	Ecuador	7,500	6.5%	27%	-0.86	0.692 (77)
	Karimunjawa	1,106	2	Indonesia	3,900	6.1%	12%	-0.50	0.593 (108)
	Wakatobi	13,436	2						
	Tubbataha	968	3	Philippines	3,300	3.8%	27%	-0.48	0.635 (97)
	Ha Long Bay	1,533	2	Vietnam	2,800	6.2%	11%	-0.56	0.566 (113)

MPA governance approach	Short name	Area (km^2)	Effectiveness*	Country	National Per Capita GDPUS$	GDP annual growth rate*	Pop. below poverty line*	State capacity*	Human development index (world ranking)*
(III) Managed primarily by local communities under collective management arrangements	Os Miñarzos	21	3	Spain	34,600	0.9%	21%	0.95	0.861 (20)
	Isla Natividad	7 (6 MPAs)	3	Mexico	14,300	1.3%	47%	-0.14	0.745 (56)
(IV) MPAs managed primarily by the private sector and/or NGOs granted with property/management rights	Great South Bay	54	2	USA	47,500	0.4%	15%	1.36	0.899 (4)
	Chumbe	0.3	4	Tanzania	1,400	7.1%	36%	-0.29	0.392 (148)
(V) No clearly recognisable effective governance framework in place	Baleia Franca	1,560	1	Brazil	10,200	5.1%	21%	0.04	0.693 (73)
	Pirajubaé	14	0						
	Cres-Lošinj	526	1	Croatia	18,400	2.4%	21%	0.38	0.765 (51)

* Figures are for 2009 except for 'Population below poverty line', which is the closest year for which a figure is available: see Chapter 5 for explanation and sources

General findings from and patterns amongst the case studies

The case studies revealed some interesting patterns in relation to the five different governance approach categories and the related metrics. These are briefly discussed in this section, drawing on Jones et al. (2013b), along with some key points concerning the role of different categories of incentives and related governance issues.

Approach I: Government-led

MPA governance under this category is characterised by having a well-established legal framework, with the MPA objectives, responsibilities of different government agencies, and rights and obligations of the public all clearly defined. Legal incentives are the key influence in most MPA-related processes, ensuring that the conservation objectives and related legal obligations are fulfilled in MPA decision-making. However, the governance framework also provides a basis for the use of a range of incentives from the other four categories, including provision for the participation of local users through collaborative approaches, albeit collaboration that is led by the state, which has the final decision-making powers. This collaboration is subject to specific legal provisions as a means of promoting transparency, equity and compliance in achieving statutory MPA obligations. The case studies in this category show that having a strong legal framework does not preclude opportunities for user participation. For example, the GBRMP rezoning process and the California MLPA initiative are widely recognised as good examples of combining stakeholder participation and scientific knowledge (Fernandes et al., 2005; Gleason et al., 2010; Olsson et al., 2008), the former through formal consultation and feedback approaches and the latter through more participative but still state-led approaches, and they are both underpinned by strong legal mandates and political leadership, which enabled coordinated and sustained efforts over relatively large spatial and temporal scales.

This study included five case studies in three countries (Australia, the UK and the US) in this governance approach category. The three countries represented have a relatively low proportion of people living below the poverty line (average 14 per cent) and high per capita GDPs (average US$41,300), state capacities (average +1.5) and HDIs (average 0.894), while the MPAs have a relatively high effectiveness (average 3) (Table 15). This approach would thus seem to be most appropriate to MEDCs with strong state-federal governance frameworks, and well-established legal and judicial systems.

Approach II: decentralised

MPA governance under this category is characterised by a sharing of authority and responsibilities between central/federal governments and lower levels of government, or between government agencies and NGOs/private entities. MPAs are governed in accordance with formal regulations and/or through partnerships and negotiations between different parties. This study included seven case studies in six countries (China, Colombia, Ecuador, Indonesia, the Philippines and Vietnam) within this governance approach category. The six countries represented all have a relatively high proportion of people living below the poverty line (average 21 per cent) and GDP annual growth rates (5.7 per cent), and relatively low per capita GDPs (average US$5,400), state capacities (average –0.54) and HDIs (average 0.638), while the MPAs have a medium effectiveness (average 1.9) (Table 15). This approach would thus seem to be characteristic of LEDCs undergoing various forms of decentralisation, where there is a degree of commitment to conserve marine biodiversity and promote sustainable fisheries but a relatively weak state capacity, hence the tendency towards decentralisation. A variety of governance incentives are employed in MPAs adopting this approach, the provision of alternative livelihoods to local communities, reinvesting tourism revenue to support both MPA management and community development, and promoting user participation in planning, monitoring and enforcement being of particular importance.

Approach III: Community-led

MPA governance under this category is characterised by local communities taking a lead in the conservation and sustainable management of marine resources. Community organisations (e.g. local fishing cooperatives) are often granted a significant level of autonomy to collectively decide the rules governing MPA management. External organisations, such as government departments and NGOs, may have an important role in enabling and reinforcing such community initiatives, and ensuring that community efforts are consistent with existing legal and policy objectives at a national or supranational level, including fisheries and biodiversity conservation obligations.

This analysis has examined two case studies in this governance approach category in two countries (Spain and Mexico). The two countries represented differ significantly in their socio-economic contexts, with a medium (24 per cent) to very high (47 per cent) proportion of the population living below the poverty line, per capita GDPs (US$) of 34,600 and 14,300, and state capacities of +0.95 and –0.14, while the MPAs have a relatively high effectiveness (3 in both MPAs) (Table 15). This approach would thus seem to be opportunistic, but effective in certain contexts, particularly when communities are empowered to develop and enforce rules for managing shared resources, subject to certain conditions related to biodiversity and resource conservation. Again, a range of incentives from different categories are applied, with economic incentives playing a particularly important role.

One of the major challenges to MPA governance in this category is that although existing governance arrangements have been successful in addressing over-exploitation of valuable marine resources in the current context, they are particularly vulnerable to changes in the wider socio-economic and political environment.

These changes include shifts in external markets, which may devalue products and services from a MPA, shifts in the political will to renew community rights to marine resources, as in the case of the Isla Natividad MPA, or national economic crises, which can undermine the national government's commitment and capacity to assist in enforcing MPA restrictions, as in the case of Os Miñarzos. Such wider-scale changes and related driving forces may significantly influence the local capacity to effectively control natural resource use.

Another concern is that the power awarded to some community organisations and groups may generate equity concerns, which may lead to the exclusion of non-elite members of a community. For example, in the Isla Natividad MPA, the Mexican constitution indicates that anyone may use resources for subsistence, but from the perspective of the fishing cooperatives this is the kind of loophole through which poaching occurs. In this case, a MPA may be an effective way for the cooperatives to ensure that even 'subsistence' use of resources like abalone is further restricted for members of the community who are not members of the cooperative, reinforcing and potentially widening local inequities. Whilst governance structures in community-based MPAs such as Isla Natividad may appear to be non-hierarchical, they can actually represent hierarchical structures based on local entitlements. Isla Natividad's communities were discussed at the MPAG workshop in terms of 'a hierarchy of wannabees', reflecting the desire amongst non-members of the co-operative to become affiliated with the cooperative in order to gain access rights to the lucrative abalone fishery.

Approach IV: Private

MPA governance under this category is characterised by non-governmental and/or private organisations taking the main responsibility for MPA management and enforcement. These organisations are often granted with permanent property rights or temporary management rights to a particular area of sea, which is managed for conservation and sustainable resource use. Such organisations generally work independently of their own volition, but often also collaborate with government bodies to enhance conservation efforts. A variety of incentives are used to steer MPAs in this category, including reinvesting profits generated from ecotourism to support MPA management and community development in a sustainable manner (Chumbe), providing environmental education to a wide range of audiences such as community members, students and tourists (Chumbe), and developing participative governance structures and processes that bring together community, government and NGO representatives in MPA initiatives (Great South Bay and Chumbe).

This study has examined two case studies in two countries (the US and Tanzania) within this governance approach category. The two countries represented are even more heterogeneous than those in Category III, with the highest and the lowest figure for per capita GDPs (US$47,500 and 1,400), state capacities of +1.36 and –0.29, and HDIs of 0.899 and 0.392, respectively. The proportion of the population living below the poverty line is also very different (15 per cent and 36 per cent).

Given these major differences between the US and Tanzanian contexts, it is particularly notable that while Great South Bay (US) has an effectiveness score of 2, Chumbe (Tanzania) has the highest effectiveness score of all the case studies, at 4 (Table 15). This indicates that such private MPAs can be highly effective, even in challenging national contexts, this being significantly attributable to the very strong leadership role of the founder of the Chumbe MPA, coupled with measures that spread the MPA's benefits to local communities. However, it must be recognised that Chumbe is also the smallest of the case study MPAs, at just 0.3km², and that small MPAs tend to be more effective (McClanahan, 1999; Mascia et al., 2010) as they pose fewer challenges in terms of their spatial scale, though this MPA also poses major challenges in terms of its national context.

Approach V: No clearly recognisable governance framework

MPA governance in this category is undermined by a lack of political will, leadership and state capacity at all levels to develop effective governance structures and processes that would support the achievement of any MPA objective, often in the face of strong driving forces. Few incentives are successfully applied to steer MPA processes and address conflicts.

This study has examined three MPAs in two countries (Brazil and Croatia) in this governance approach category. The two countries represented have a medium proportion of the population living below the poverty line (21 per cent), medium per capita GDPs (US$) of 10,300 and 18,600 and state capacities of +0.04 and 0.38, while the MPAs have a low effectiveness (0–1). It should be noted that strong driving forces are not unique to MPAs in this category, in that several other MPAs, particularly those under governance approach II, are also facing similar conflicts and challenges to achieving conservation objectives. Such conflicts can be mitigated and reduced through the use of different incentives, as shown in other case studies, but considerable leadership and commitments, be it from the state, NGOs, the private sector and/or communities, particularly the former, must underpin the use of such incentives. Compared to MPAs in other governance approach categories, the limited use of economic and legal incentives in MPAs in this category is most notable, as they are important elements of the MPA governance framework in the other categories examined, providing the 'carrot' and/or 'stick' that are needed to steer MPA governance.

Driving forces

The MPA case studies were found to be exposed to various driving forces that could raise challenges for their governance frameworks, potentially steering human behaviour in more conflicting directions, i.e. various forces that could drive user activities in such a way as to undermine the effectiveness of MPAs in achieving their objectives. Some of the main driving forces are outlined in this section, along with some examples of the challenges raised by these driving forces in relation to specific case studies.

Global fish markets

This is related to the increasing reach and demand of global fish markets, for growing and increasingly affluent human populations, coupled with the increasing reach and effort capacity of fishing vessels, e.g. Galápagos – lucrative opportunities for local and incoming vessels to export to Asian fish markets, particularly for sea cucumbers, shark fins and lobsters; Seaflower – increasing fishing pressure, mainly from incoming industrial trawlers, for lobsters for export to US markets; Isla Natividad – lucrative abalone export market stimulating harvesting effort as well as efforts to restrict access to fisheries; Wash – increasing demand from the Netherlands for mussels, promoting increased efforts to harvest and ranch mussels. IUU fishing is a particular problem in this respect, often involving organised crime due to the lucrative nature of fish markets and the related opportunities for drug smuggling, people trafficking, forced labour, etc. IUU fishing vessels will often attempt to breach any fishing restrictions, including for MPAs, and it can be very challenging to control such vessels. Legal and legitimate fishing vessels can also raise challenges, however, as there is often a political reluctance to restrict potentially lucrative fishing operations. Such driving forces raise major challenges for the objective of restoring and conserving fish stocks as they tend to stimulate demand for fish and thereby efforts to harvest fish, directly impacting the MPA's fish stocks. They also particularly raise challenges for the objectives of restoring marine ecosystems, and protecting rare and vulnerable habitats and species, as the direct and indirect impacts of fishing can lead to trophic cascades, habitat impacts, non-target population depletions, etc.

Tourism

This is related to the increasing reach and numbers of tourists, as the global corporate tourism market expands, both for large-scale mass holidays and more elite luxury holidays, e.g. Karimunjawa and Sanya – rapid growth of domestic and regional tourism; Galápagos and Wakatobi – rapid growth of international tourism, including growth of more damaging 'adventure' tourism in Galápagos; Cres-Lošinj – increasing numbers of recreational motorboats that could impact dolphin populations. Tourism can have direct and indirect impacts, as discussed in Chapter 2, raising challenges for the objective of controlling the impacts of tourism and recreation itself and thereby for the objectives of restoring marine ecosystems, and protecting rare and vulnerable habitats.

Coastal development

This is related to the pressures for economic development through large-scale coastal development, urbanisation and infrastructure projects, e.g. Sanya – rapid expansion of hotels for mass tourism; GBRMP – proposals for port expansion to serve shale gas and coal sector, and intensification of agriculture in coastal catchment; Cres-Lošinj – development of large marina; Pirajubaé – reclamation and dredging in the bay to build a highway to an airport; Ha Long Bay – pollution from nearby urban centres. The prioritisation of economic development can lead national and/or local governments to collaborate with the private sector in promoting such developments, as was the case in some of these examples. Developments can undermine the achievement of objectives to restore and conserve marine ecosystems and fish stocks and to protect rare and vulnerable habitats and species. The objective of contributing to marine spatial planning can support or challenge such developments, depending on whether an integrated use or ecosystem-based perspective on MSP is adopted, as was discussed in Chapter 2.

Poverty

This is related to the dependence of local human populations on marine resources for subsistence, e.g. Chumbe – case study with the lowest per capita GDP and HDI, leading to a high dependence on marine resources, other case studies such as Karimunjawa, Wakatobi, Tubbataha and Ha Long Bay facing similar challenges. Where local communities are dependent on marine resources, this can lead to myriad small-scale uses that can collectively undermine the achievement of objectives to restore and conserve marine ecosystems and fish stocks and to protect rare and vulnerable habitats and species. Furthermore, restrictions on these uses on which people are dependent, in order to promote the achievement of such objectives, can raise major equity issues.

Aspirations

This is related to the increasing and legitimate desire of local populations, which may themselves be increasing through population growth, to not only feed themselves and their families on a subsistence basis, but also to improve their material standard of living and their prospects. This is particularly the case given that many people increasingly aspire, including through media exposure and other aspects of globalisation, to a more secure, comfortable, consumerist and technological 'western' lifestyle. This can lead to people aiming to expand subsistence activities into small-scale and larger scale commercial activities, including in relation to the driving forces of global fish markets, tourism and other developments, e.g. Galápagos, where the opportunities provided by access to global fish and tourism markets are stimulating impacting activities through such aspirations. Such aspirations need to be recognised and, as far as is feasible and appropriate, provided for in MPA governance.

Migration

This is related to the enhanced mobility of people and thereby the increasing potential for migration to coastal areas, where economic development and subsistence opportunities tend to be relatively good, e.g. Galápagos – average income on the main island is twice as high as on mainland Ecuador due to tourism and fishing opportunities, creating a major 'pull' factor for migration and leading to a 'gold rush' and 'mainland-isation' culture; Seaflower – migration from mainland Colombia for similar reasons. Migration increases the pressure on fisheries and promotes tourism and other development pressures. Many migrants arrive in a state of poverty, increasing subsistence pressures, and they often aspire to increase their standard of living, this being one of the reasons they were 'pulled' to the MPA, increasing fishing, tourism and other development pressures. Poverty, aspirations and migration can combine to undermine the achievement of many MPA objectives, particularly to restore and conserve marine ecosystems and fish stocks, to protect rare and vulnerable habitats and species, to control the impacts of tourism and recreation, and to maintain traditional uses.

Combining to perturb governance and undermine effectiveness

Such driving forces, acting in combination, are a major and increasing challenge for MPA governance, as they represent forces that can combine to perturb, disrupt and, ultimately, collapse governance systems and thereby undermine effectiveness in achieving MPA objectives. Declines in marine resources and of the health of ecosystems as a result of such driving forces have been well documented in the literature and observed in several of the case studies examined in this study, particularly Galápagos and Sanya. Furthermore, the findings from the case studies support the argument that such driving forces not only lead to biodiversity loss and ecosystem degradation, but may also threaten traditional livelihoods and lead to the inequitable distribution of costs and benefits associated with MPAs. For example, the increasing reach and influence of industrial fishing fleets and corporate tourism industries have resulted in the marginalisation of traditional resource users and an inequitable distribution of benefits in several case studies, e.g. Galápagos, Seaflower and Sanya.

The next chapter will focus on detailed discussions on how individual incentives have been used in different case studies to withstand the perturbing effects of such driving forces in order to contribute to governance frameworks that are more robust and effective.

Chapter 7

Incentives for effectiveness

Introduction

In the context of this book, incentives are defined as 'particular institutions that are instrumentally designed to encourage actors to choose to behave in a manner that provides for certain strategic policy outcomes, particularly MPA objectives, to be achieved'. In keeping with this focus on incentives, which form the central element of the MPA governance empirical framework, and in keeping with the co-evolutionary hierarchical governance concept discussed in Chapter 4, a new definition of governance is proposed – 'Steering human behaviour through combinations of state, market and civil society approaches in order to achieve strategic objectives'. As was highlighted in Chapter 5, the 'bottom line' for the analysis of MPA governance is whether the incentives that are being used are leading to effectiveness, i.e. are activities being appropriately controlled and restricted in order to reduce their impacts and thereby ameliorate or resolve conflicts between certain uses and the achievement of MPA objectives? The analysis of incentives is therefore directional in that the focus is always on effectiveness in achieving MPA objectives.

Of course, the behaviour of people is influenced by many other objectives and related incentives, e.g. to increase their economic wealth by increasing the income they gain from MPA related activities, but these are considered in this analysis in terms of the driving forces that tend to encourage behaviour that could exacerbate conflicts and thereby undermine the achievement of MPA objectives, such driving forces being what MPA incentives ultimately have to address. Many governance analysts, particularly those inclined towards promoting self-governance, will be resistant to the instrumental nature of the above governance definition, given its directionality in focusing on steering human behaviour towards the achievement of MPA objectives (detailed in Chapter 2) and withstanding driving forces that could divert the MPA from this course. However, accepting that MPAs are essentially defined by such objectives, these being what distinguish MPAs from wider sea areas, and that we cannot actually govern marine ecosystems *per se*, it is logical to focus on MPA governance in terms of steering human behaviour as a means of achieving such objectives. Trying to focus on governance that is not directional and has no steer reminds me of attempts to sail a dinghy without the rudder in position: the wind forces the dinghy's direction and you have little control over its course or angle to the wind, usually resulting in grounding, collision or capsize.

Incentives used to withstand driving forces

In the MPAG case studies, a variety of different incentives has been applied to support the effective achievement of various MPA objectives in the face of the driving forces outlined in the previous chapter. These incentives represent different sources of 'steer' in MPA governance (Table 8), promoting decisions and behavioural change in a way that provides for the achievement of a given MPA's objectives. The challenges raised by the driving forces can arguably only be addressed through a co-evolutionary hierarchical governance approach, involving a combination of incentives. If such driving forces are not addressed, the effectiveness of the MPA will be undermined through the perturbation and potential collapse of the governance framework. The five categories of incentives (listed in alphabetical order) on which this analysis of MPA governance is focused were introduced in Chapter 5 (Table 11) and this chapter will focus on the 36 incentives (listed in Table 12) that form the elements of the governance framework for MPAs, including an overview and examples of how they have been applied in relevant case studies.

Economic incentives

Neoliberalism as a context

The pervasive nature of market forces and related neoliberal tendencies, discussed in Chapter 4, are discussed further here in terms of economic incentives. Though not all of these incentives are necessarily linked to neoliberalism, it is worth briefly considering this concept as it is of considerable significance in governance debates, being a key argument for market steer (Table 8). Neoliberalism is a somewhat loose and sometimes disputed term, in that there are many perspectives on it and it is difficult to define, but it is broadly accepted as involving the *laissez-faire* shrinking of the state and reduction of state steer through deregulation, and the unleashing of market forces, including privatisation and marketisation. Organisations such as the World Bank, regional development banks and the International Monetary Fund are considered by many as promoters of neoliberalism. However, this concept and its practice, through approaches such as the allocation of property rights to natural resources and the focus on the economic value of ecosystem services, are also widely criticised, e.g. McCarthy and Prudham (2004), as will be more specifically considered below in relation to some of the economic incentives.

At this point it is worth noting that the originator of the metaphor that the 'invisible hand' of market forces amongst self-interested individuals can efficiently guide decisions to achieve societally optimum outcomes, Adam Smith, recognised that state intervention is required in some contexts (Stein, 1994), particularly public goods such as the environment, which Adam Smith discussed in relation to the challenges of keeping streets clean and the need for state intervention in such contexts. Furthermore, the recent global collapse of many economic systems is attributed by some to a lack of state regulation, which allowed short-termism and self-interest to prevail in many markets, though this is disputed (Anon., 2008). This might be considered as a warning concerning the potential collapse of ecosystems should such market forces be unleashed in biodiversity and natural resource conservation governance through a lack of regulation of natural resource property rights and markets, and a related lack of state steer.

In a similar vein, Young (2010) notes that markets represent the sort of 'preconceived and typically simplistic remedies' that are often prescribed but that there is little reason to expect that markets alone will suffice to maintain the resilience of social-ecological systems, whilst Sandel's (2012) discussions highlight that markets have moral limits, in that they are not able to protect the 'good things in life' (p.10) that really matter to society. These critiques, however, do not preclude the useful role that market incentives can have as tools[17] for implementing decisions taken through other more societally accountable and inclusive means, i.e. market forces can be harnessed rather than unleashed, provided that other incentives are employed to ensure that the things that really matter, such as the health of our seas and the equity of our society, are protected.

Payments for marine ecosystem services (PESs)

This incentive is focused on direct payments for the flow of ecosystems services provided by a given MPA through formal markets. Marine PESs are currently focused mainly on blue carbon initiatives, which were discussed in Chapter 4 as an example of the pervasive nature of market forces. Blue carbon initiatives are the marine equivalent of terrestrial PES schemes such as that for reducing emissions from deforestation and degradation, conserving and enhancing forest carbon stocks, and sustainably managing forests (REDD+), in that they are a means of providing payments to sustain the use of mangrove, tidal marsh and seagrass habitats as carbon sinks to mitigate climate change. Blue carbon initiatives are still in their infancy, mainly confined to preliminary research, demonstration and pilot projects around the world. Whilst blue carbon initiatives have not yet matured into actual schemes connected to formal carbon trading markets, there are aspirations that they will do so, along with other marine PESs, and that MPAs will provide an ideal vehicle for such initiatives (Lau, 2012; Rees et al., 2012).

REDD+ initiatives are much further developed, though they still essentially represent large-scale PES experiments, and there are many concerns about the practice and implications of such PES initiatives (Redford and Adams, 2009; Corbera and Schroeder, 2011; Corbera, 2012). Such concerns include the potential to reduce biodiversity to only those species and habitats that provide valuable ecosystem services, to displace harvesting and other extractive activities to other locations through 'leakage', to reduce wider preservationist and ecocentric values and related justifications for conservation to narrower utilitarian values and justifications (Table 6), and to lead to inequities through PESs not being shared with and amongst local people, who often bear opportunity costs related to protection through PES initiatives. There are also related concerns that PESs could lead to an 'ecosystem services curse' (Kronenberg and Hubacek, 2013).

None of the case studies identified PESs as an incentive that was being used in any of the 20 MPAs, nor were PESs identified by any of the case studies as being an incentive that was particularly needed. This indicates that the major interest in proposals and trials for PES initiatives such as blue carbon have not yet been translated into interest at a MPA specific level, other than in specific trial and research contexts. It is also important to recognise that, in addition to the concerns such as those discussed above, many marine habitats are not major carbon sinks so would not qualify for such PESs, limiting the potential applicability of PESs, unless other marine ecosystem services besides climate change mitigation are included.

Assigning property rights

This incentive is focused on assigning property rights for certain marine areas and fisheries to appropriate groups of people to promote ownership, stewardship and rational self-interest in sustainable exploitation. As with PESs, this incentive is consistent with neoliberal ideologies and theories, the World Bank being an advocate of assigning property rights to fish stocks to address fisheries management challenges (Arnason et al., 2008). Such approaches are also supported by some fisheries scientists and marine ecologists (e.g. Hilborn et al., 2005; Fujita and Bonzon, 2005; Costello et al., 2008) on the basis that they protect fish stocks, habitats and the communities that depend on them. The assignation of property rights is also argued for by neo-institutionalists, especially when couched in terms of assigning tenure to the local user community (Hayes and Ostrom, 2005), this approach in itself being argued to be more effective than protected area designation. These arguments are similar to those of others, such as Hilborn et al. (2004), that MPAs do not address the root causes of fisheries mismanagement, i.e. weak incentives due to lack of resource ownership. Proponents of more community-based governance similarly recommend strengthening customary marine 'tenure' systems (Asafu-Adjaye, 2000), 'enclosure' through the assignation of individual transferable quotas (ITQs) to a particular community of fishers for a particular area (McCay et al., 2008), and the use of 'territorial user rights in fisheries' (TURFS) (Gelcich et al., 2008). Whilst such approaches are primarily aimed at achieving sustainable natural resource use, areas managed by fishers through such rights can show add-on biodiversity conservation benefits (Gelcich et al., 2008), making them *de facto* MPAs and arguably demonstrating that rights-based approaches to fisheries management are more effective than many actual MPAs.

There are, however, critiques that call the enthusiasm for a rights-based approach to fisheries into question. Whilst biodiversity conservation benefits such as those reported above can be delivered through the assignation of property rights, there is no guarantee that they will be, nor can it be taken for granted that they will be. This is particularly the case given the divergences between sustainable resource use and biodiversity conservation objectives (Chapter 3). Swan and Gréboval (2005) report several case studies that reveal that such approaches can still result in overfishing, especially in the face of uncertainty and conflicts, and that success depends on addressing allocation dilemmas from both a human and ecosystem perspective. They also stress that regulating access and dealing with displaced fishermen are particularly important problems that must be overcome, and that the race-to-fish can become a race-for-rights, which brings its own problems. Beddington et al. (2007) stress that 'the simple creation of rights-based incentives does not automatically deal with ecosystem problems' and that MPAs therefore have an essential role in addressing the impacts of fishing in relation to ecosystem conservation priorities. Symes (2000) similarly argues that rights-based approaches to fisheries management would be unlikely to address wider societal concerns about the impacts of fishing on the productivity, diversity, integrity and service provision functions of marine ecosystems and that ecosystem-based approaches to address these concerns, such as no-take MPAs, 'are not the kinds of actions that can reasonably be left to the fishing industry to formulate and implement'.

Such arguments aside, several of the case studies illustrate how property rights can be employed as an incentive (Box 2).

Box 2 Case study examples of assigning property rights

Sanya

Sea user rights allocated to tourism developers to promote confidence and investment.

Galápagos

Rights to operate tourism vessels ('cupos') allocated to Galápagos residents.

Os Miñarzos

TURFs assigned to local fishermen to promote collective management.

Isla Natividad

Harvesting concessions allocated to the local fishermen's cooperative.

Great South Bay

Use rights held by NGO were employed to end shellfish harvesting in the MPA.

Measures to reduce the leakage of benefits

This incentive is focused on measures to reduce the 'leakage' away from local people of the economic benefits arising from compatible activities in the MPA, including measures to promote the fair distribution of such benefits amongst local people. Such benefits are often related to the successful fulfilment of certain MPA objectives, such as increased harvests on the edge of no-take areas or income from sustainable tourism. The rationale behind this incentive is that local people who may bear some of the opportunity costs associated with restrictions in access to natural resources in a MPA should be able to yield a sufficient proportion of the benefits to offset these costs and provide for their needs and development. If such benefits are gained by incoming or neighbouring fishermen, incoming tourism development interests, etc., this undermines both the equity of the MPA, through the unjust distribution of costs and benefits, and potentially its effectiveness, as economically marginalised local people may have to resort to breaching MPA regulations in order to provide for their needs and development. Several case studies attempted to reduce such leakage (Box 3).

Box 3 Case study examples of measures to reduce the leakage of economic benefits

Seaflower

Several artisanal fishing zones for exclusive use by traditional fishers are located adjacent to NTZs to ensure that the benefits of any export/spill-over go mainly to local fishermen. Jobs related to the MPA are only given to island residents.

Galápagos

Excluding incoming vessels by confining access to the MPA's fishing grounds to Galápagos vessels. Only allocating rights to operate tourism vessels to Galápagos vessels.

Tubbataha

Residents on islands in the municipality of Cagayancillo are given preferential consideration in employment opportunities in the MPA. Levy of entry fees for each tourist covers some MPA costs but these may have to be increased or the MPA authority may start operating tourism vessels itself in order to minimise the leakage of benefits to external vessel operators.

Isla Natividad

'Intrusion' by incoming diving and surfing operators restricted so that locals can capture the benefits by running such activities themselves. Fisheries harvesting concessions are allocated exclusively to local fishing cooperative members.

Promoting profitable and sustainable fisheries

This was previously discussed in relation to the objective of restoring and conserving fish stocks (Chapter 2), but it is discussed here as an economic incentive as such fisheries are clearly a MPA-related benefit of relevance to fishermen. The achievement of this objective generally involves promoting profitable and sustainable fisheries by restricting harvests to MSY levels and restricting or banning fishing methods that can impact conservation features. It can also involve providing a refuge for marine organisms in NTZs in order to safeguard and enhance harvests in adjacent fishing grounds through spill-over/export, insurance against uncertainty, increased resilience, etc., though there are divergent views amongst scientists as to the necessity of such approaches from the perspective of promoting profitable and sustainable fisheries, as was discussed in Chapter 3.

Most of the case studies included the objective of restoring and conserving fish stocks and thereby also applied this incentive, with the exception of the Darwin Mounds, Sanya and Ha Long Bay, where this did not seem to be a priority. Some case studies were entirely no-take MPAs, e.g. Tubbataha, Great South Bay and Chumbe, or included NTZs, e.g. GBRMP, California, Seaflower, Karimunjawa and Isla Natividad, and these have led to some recognised benefits for surrounding fisheries that represent an incentive in this respect (Box 4).

Box 4 Case study examples of spill-over/export benefits

GBRMP

Recruitment in commercial fish stocks in the vicinity of NTZs significantly enhanced by larval export (Harrison et al., 2012).

Tubbataha

Increased CPUE in fisheries around the MPA, more than tripled in some cases.

Isla Natividad

Slightly greater body sizes and densities of pink abalone populations in NTZs enhance recruitment in the vicinity of the NTZs, particularly after a climate driven hypoxia episode, as unexploited pink abalone populations seemed to be more resilient to hypoxia (Micheli et al., 2012).

Great South Bay

Major re-stocking programme has increased the density of hard clams, which has significantly benefited surrounding hard clam fisheries through the export of propagules.

Chumbe

Biomass of commercially exploited fish inside MPA has increased and yields in vicinity of MPA reported by fishermen to have increased.

Promoting green marketing

This incentive is focused on promoting the 'green marketing' of tourism, fish, etc. products from the MPA in order to increase incomes for local people through such price premiums, by way of providing benefits to compensate for opportunity costs related to restrictions on harvesting or otherwise using MPA resources. This involves not only green marketing to promote demand for products, but also to increase the price per unit of product that users receive, as a focus only on the former may paradoxically increase the impacts of resource extraction. This can also involve more directly connecting local MPA users to external markets, as was discussed in Chapter 4, so that they get a greater proportion of the sale value, rather than intermediate merchants and agents making more profit than local users, though it must again be ensured that this does not lead to an increase in demand and thereby in impacts.

One of the key means of implementing this incentive is to develop environmental accreditation schemes, by which the products from MPAs can be marketed as being 'green' and thereby attract a premium price and/or promote sales, as ethical consumers strive to source sustainable products. Whilst such schemes for fisheries (Jacquet et al., 2009) and tourism (Haaland and Aas, 2010) have been critiqued on the grounds that they do not necessarily require actual sustainability and lack fully independent environmental auditing, i.e. they represent 'greenwashing', there is the potential to improve such green marketing schemes by explicitly aligning them with MPAs that are focused on effectively and equitably achieving specific MPA objectives, particularly that of controlling the impacts of tourism (Chapter 2), in order to provide for the achievement of other MPA objectives. As well as promoting green marketing, such approaches could be effective in that they could require users to adhere to strict environmental standards that are specifically tailored to a given MPA, provided these requirements are enforced. Several case studies provided examples of how this incentive could be beneficial in relation to the green marketing of MPA resources (Box 5).

Box 5 Case study examples of promoting green marketing

GBRMP

Tourism operators with very high environmental standards, which are certified by an independent eco-certification programme, are promoted by GBRMP Authority, the majority of tourists being carried by such operators.

NE Kent

In the face of the decline of traditional seaside tourism in this area, the MPA has been used as a means of promoting the marketing of the coastline as an ecotourism destination.

Seaflower

Developing markets for seafood sourced from the MPA as a green sustainable product.

Galápagos

Potential for the green marketing of tourism and fish products is declining as the conservation status of fish populations and wider biodiversity declines, market demand for both being so large as to potentially be a driving force of depletions.

Tubbataha

Promoting the MPA as a premier boat-based diving location, capitalising on its World Heritage Site status including charging an entry fee for dive tourism vessels as a means of financially supporting the MPA.

Os Miñarzos

The MPA has been used to promote green tourism through activities such as underwater photography competitions, tourist trips on working fishing boats and seafood cookery workshops.

Chumbe

The MPA has won many international awards as a premium and effective ecotourism destination, which has helped in advertising and in ensuring a sustainable flow of tourism income on which the funding of the MPA depends.

Isla Natividad

Whilst accreditation by the Marine Stewardship Council and access to new lobster markets did not prove to be beneficial, abalones are marketed as coming from this sustainable source.

Promoting alternative livelihoods

This incentive is focused on offsetting opportunity costs incurred by users as a result of MPA restrictions on using marine resources by providing alternative livelihoods and small-scale commercial activities that are compatible with the MPA's objectives, bearing in mind that many users of MPAs want more than subsistence livelihoods. It can also be used to promote compliance with MPA conservation measures, given that food and income insecurity are recognised as key drivers of non-compliance with MPA regulations (Peterson and Stead, 2010). This is particularly important where there is an emphasis on voluntary compliance with MPA restrictions and where the marine resource dependence of local people means that it is not equitable or realistic to restrict access to marine resources without considering alternative livelihoods.

There are concerns that such alternative livelihoods may fail to be economically sustainable or to actually compensate for losses of access to MPA resources, may not be acceptable to locals as an alternative way of life, given that fishing can be a valued way of life, or may not actually be compatible with the MPA's objectives, in that they promote activities that can impact the MPA's conservation features. Such concerns are compounded by the tendency that the actual effectiveness of the success of alternative livelihoods is rarely evaluated. However, promoting alternative livelihoods remains an important priority for many MPAs, particularly those in LEDCs where local people have a high dependence on marine resources. Several of the case studies illustrate the application of this incentive through initiatives to try and promote alternative livelihoods (Box 6).

Box 6 Case study examples of promoting alternative livelihoods

Sanya

More than 1400 jobs created through the rapid development of tourism.

Seaflower

Pilot projects through seaweed, breadfruit and iguana farming, marine tourism led by artisanal fishers, and a community-run mangrove park.

Galápagos

Develop the capacity for local people to process, market and export seafood products in order to maximise their income from fishing and thereby reduce fishing pressure.

Karimunjawa

Micro-finance schemes, establishment of mariculture zones and technical support by NGOs to promote the development of seaweed, clam and fish farming for domestic markets, whilst also controlling and minimising the related environmental impacts.

Tubbataha

Micro-finance schemes by NGOs to promote alternative livelihoods, such as running a homestay or eatery, trading in household commodities, weaving mats and producing coconut vinegar. Seaweed farming was attempted but was not viable.

Chumbe

The ecotourism resort has an exceptionally high staff–tourist ratio in order to promote jobs for local people, as well as providing livelihoods related to tourism, e.g. food and handicrafts market on island, provision of craftsmen and building materials for resort maintenance, outsourcing of road and boat transport.

Providing compensation

This incentive is focused on providing fair economic compensation for those users who carry costs as a result of restrictions on their activities that cannot reasonably be offset through alternative livelihood and development opportunities. This can be at a national scale, whereby the government receives compensation from a trust fund, to which international donors can contribute, for the loss of revenues as a result of no longer selling fishing licences to foreign vessels due to the designation of NTZs, as is the case with the Phoenix Islands Protected Area 'reverse fishing licence' scheme (Anon., 2013). The more money is donated to the fund, the more of the MPA is designated as NTZ. Usually, however, such compensation is paid by national governments to individual fishermen and related economic interests. There are concerns that, on the one hand, such compensation never reaches all excluded users or can never fully compensate local users for the costs incurred by loss of access to marine resources through MPA restrictions, or on the other hand, that compensation costs can spiral out of reasonable proportion,[18] as illustrated by the three case studies that applied this incentive (Box 7).

Box 7 Case study examples of providing compensation

GBRMP

A structural adjustment package (SAP) was offered for fishermen, their employees and other businesses that were negatively impacted by the expansion of NTZs in the MPA rezoning. An initial budget of AUS$10 million was allocated but the total amount of compensation paid spiralled out of control to $250 million (Macintosh et al., 2010): much more than the rest of the rezoning budget ($10–15 million).

Wakatobi

Dive tourism operator pays equivalent of US$500 per month to each of the 17 villages affected by the exclusion of fishermen under the 'reef leasing' scheme, but none of this goes directly to the affected subsistence fishing communities, who were marginalised from the negotiations.

Tubbataha

Ten per cent of the tourism entry fees are paid to the municipal government to fund an alternative livelihoods programme.

Reinvesting MPA income in local infrastructure

This incentive aims to ensure that some of the MPA income flows to the local government or local community organisations for reinvestment to develop local facilities (schools, medical care, etc.) and infrastructure (roads and other transport links, electricity, water, etc.). Several of the case studies illustrate the application of this incentive (Box 8).

Box 8 Case study examples of infrastructure reinvestment

Sanya

Tourism companies funded improvements in public infrastructure, such as schools and roads in the village of Ximao.

Karimunjawa

Improvement of local infrastructure to support development of tourism and mariculture industries in MPA zones where this is identified as a priority.

Wakatobi

Repairs to public buildings and transport infrastructure are funded by compensation provided by dive tourism operator under the 'reef leasing' scheme.

Tubbataha

Seven to ten per cent of the tourism entry fees are used to fund infrastructure improvements, such as new roads and the improvement of public facilities on islands in the municipality of Cagayancillo. Fees also support the better management of fisheries outside the MPA for Cagayancillo fishermen.

Ha Long Bay

Tourism fee income reinvested to improve living conditions and provide educational facilities on the floating village. Also reinvested, along with major development bank funding, to improve industrial and domestic waste water treatment in the city to reduce land-based pollution.

Chumbe

Tourism income used to provide environmental education facilities and opportunities. Warden patrol boat also serves as a rescue boat for local fishermen outside the MPA.

Isla Natividad

Some of the income from the fishing cooperatives is used to fund improvements in electricity and water supply infrastructure. Whilst this is mainly targeted on shellfish processing plants, houses and other buildings in the vicinity also benefit.

Ensuring sufficient state funding

This incentive is focused on ensuring that a sufficient degree of state funding is available to support the governance of the MPA, particularly in relation to the enforcement of legal incentives and contributing funds for the other economic incentives discussed in this section. In keeping with the need to combine top-down and bottom-up approaches through the co-evolutionary hierarchical governance concept, it is important to achieve a balance between providing sufficient state funding to support the achievement of strategic objectives whilst ensuring that such funding does not allow the state to undermine the role of participative incentives.

A large proportion of the funding for most of the case studies was provided by the state, particularly those categorised as being governed primarily by the state (MPA Governance Approach I, Table 15). The MPA case studies categorised as decentralised (Approach II) or managed primarily by local communities (Approach III) generally gained a larger proportion of their funding from NGO or private sector sources, but state funding was still important, including from the GEF and development banks, often still representing the majority of the funding. Charging 'user fees' for tourists is also an important source of funding in several case studies in these categories, e.g. Seaflower, Galápagos, Wakatobi, Tubbataha and Ha Long Bay, though this generally represents only a small proportion of the total funding for these MPAs. The two 'private sector' MPAs (Approach IV) received very little state funding, Great South Bay being NGO funded and Chumbe being funded by private investment and tourism income, the latter actually contributing significant funding to the state in the form of a tourism tax. One of the main problems with the case studies with no clearly recognised governance framework (Approach V) was a lack of state funding and an underlying lack of political will for ensuring the effectiveness of these MPAs. Whilst many MPAs benefit from contributions by NGOs, as is discussed below, it is generally state funding and underlying political will to support the enforcement of MPA restrictions that leads to effectiveness.

Provision of NGO and private sector funding

This incentive aims to secure NGO (including private and charitable foundations) and private sector funding through various means to support the governance of the MPA, whilst ensuring that such funders cannot 'capture' MPA governance through an inappropriate degree and type of influence. Such funding can be a very important supplement for the funding from the state discussed above, and even a substitute in some cases, as is illustrated by several of the case studies (Box 9).

Box 9 Case study examples of provision of NGO and private sector funding

GBRMP

The Great Barrier Reef Foundation funds research to inform management decisions and promote effectiveness, particularly in relation to resilience to climate change, by engaging with private sector companies to gain their financial support for such research.

California

The process to collaboratively design the MPAs was largely funded by the Resource Legacy Funds Foundation (RLFF), which was also influential in lobbying for the Marine Life Protection Act under which these MPAs are designated. The RLFF brokers and coordinates funding from other large foundations. One of the main challenges is the funding of the enforcement and monitoring of MPAs now that they have been designated, as state funding is limited by the financial crisis, the aim being that the RLFF will be able to continue to source funding to support MPA implementation from other foundations.

Sanya

Tourism developers pay for most of the enforcement of the MPA restrictions by covering the salary costs of 16 wardens as well as employing guards at most hotels to ensure tourists do not breach coral reef protection restrictions.

Galápagos

Much of the research to inform the management of the MPA is funded by the Charles Darwin Foundation, which receives its support from other foundations and donors. Many of the initiatives to support the governance of the MPA through various incentives are NGO funded.

Karimunjawa

An international (Wildlife Conservation Society) and a local NGO (Taka) have been working in partnership and have contributed to the funding for the MPA, along with support from other NGOs for specific initiatives in relation to various incentives.

Wakatobi

NGO and dive tourism operators have provided considerable funding, though there are concerns about the risks of institutional capture related to tourism developments coupled with the vulnerability of the MPA to the reduction of NGO funding.

Tubbataha

Nearly 90 per cent of the MPA funding comes from NGOs as the tourism entry fees are not sufficient to cover the running costs.

Os Miñarzos

WWF funded and organised the initial biological survey and monitoring of the MPA, as well as other activities such as workshops with fishers, social surveys, campaigns of public awareness and running a volunteering programme.

Great South Bay

The MPA is entirely funded by The Nature Conservancy, including the purchase of the area.

Chumbe

The MPA was originally privately funded by a lead individual as NGOs considered such a venture to be too politically and economically risky, the operational funding being entirely derived from ecotourism income.

Cres-Lošinj

The MPA has been almost entirely driven and funded by the local NGO Blue World, which was established specifically for this initiative. They gain their funding from paying volunteers, corporate donations and a European Commission programme.

Isla Natividad

The Mexican NGO, COBI, worked in partnership with the fishing cooperative, providing technical and financial support for the MPA, including brokering additional funding from another foundation. A California NGO fund and support the MPA monitoring programme.

Interpretative incentives

Introduction: interpretative enforcement

These incentives are focused on promoting awareness of the conservation features of the MPA, the related objectives for conserving them and the regulations for achieving them, and building support for related measures. It draws on the notion that comprehension is a prerequisite of cooperation, i.e. if users of a MPA comprehend and understand how their activities are impacting the MPA and the potential benefits of controlling these impacts, they are more likely to cooperate with restrictions. In an entirely top-down governance system, comprehension may not be considered as a prerequisite for cooperation, but promoting such comprehension through the use of interpretative incentives is considered to be an important means of promoting cooperation in the context of the co-evolutionary hierarchical governance concept, in which top-down and bottom-up approaches are combined. This focus on interpretative incentive was also inspired by NOOA MPA veteran, Billy Causey's recognition 'that there has to be some capacity to nudge people in the direction of compliance through "interpretive enforcement"' (Causey, 1995, 120).

Raising awareness

This incentive involves using the media, champions and various interpretative approaches to overcome 'out of sight, out of mind' hurdles (discussed in Chapter 3) by raising the awareness of users, local people, relevant authority officers, politicians, etc. about the aesthetic values, ecological importance and vulnerability of marine ecosystems in terms of the species, habitats and 'landscapes' of the MPA, and the related benefits of conserving the MPA.

Various approaches and media are used for raising such awareness, such as leaflets, web pages, posters, notice boards, interpretation centres, smart phone/ tablet apps, newspaper/magazine articles, TV features, newsletters, school education programmes, community events and presentations to specific user groups. These are all well established techniques in the field of environmental education and awareness-raising, and all of the MPA case studies used various combinations and adaptations of these techniques as appropriate to a given context. Karimunjawa and

Tubbataha adopted the 'take pride' approach, whereby the NGO RARE runs a grassroots campaign, employing social marketing techniques developed from the commercial field, to sell the MPA and build support for it – 'inspiring conservation' (www.rareconservation.org). Other case studies, such as the GBRMP, US NMSs and Tubbataha used publicly recognised 'champions', such as sports stars and TV personalities, to promote awareness and build support, the GBRMP employing various champions to enthuse support, with the use of the slogan 'let's keep it great' in a media campaign, including TV advertisements.

The GBRMP also demonstrates how raising public awareness can be instrumental in promoting political will for the more effective governance of MPAs. Having developed public support for the conservation of the GBRMP through this campaign, the GBRMP Authority then commissioned a full public opinion survey to quantitatively demonstrate the wide extent of public support for the more effective conservation of the GBRMP. They then took these opinion poll findings to politicians, which served as a mandate from the electorate and thereby helped generate the political will that led to the improved conservation through a re-zoning initiative.

It is clear from the fact that every single case study employed such awareness-raising incentives that they are a vital element of MPA governance frameworks as a key means of non-coercively steering human behaviour by nurturing support for the achievement of MPA objectives. Such awareness-raising also supports numerous incentives from the various categories, e.g. better informed and more constructive participation of users in MPA governance, and promoting cooperation and compliance with legal restrictions.

Promoting recognition of regulations and restrictions

Clearly, it is only feasible for people to choose to cooperate or accept the need to comply with MPA restrictions on certain activities in particular marine areas if they are aware of such restrictions. This incentive thereby builds on the previously discussed incentive by more specifically making people aware of the details of a given MPA, particularly its boundaries, including zones within it, and the restrictions that are in place across the MPA or in particular zones. Contrary to the legal doctrine that 'ignorance is no excuse in the eyes of the law', making MPA users aware of restrictions is often a prerequisite for successful prosecution if such restrictions are breached. Given the links with other perspectives, however, the primary aim of this incentive is to promote voluntary cooperation, though its role in supporting prosecutions of those who breach such restrictions can be an important means of avoiding a downward spiral in cooperation, as those who choose to cooperate need to know that those that do not are penalised.

Promoting recognition of benefits

This incentive is more specifically focused on promoting recognition of the potential resource benefits of the MPA in terms of spillover/export benefits for wider fisheries (such as those outlined in Box 4), insurance in the face of uncertainty, resilience against climate change, etc., whilst being realistic about such potential benefits and not 'over-selling' them, recognising that promoting MPAs as win-win solutions for both marine biodiversity and fisheries conservation could be considered to be inappropriate if not problematic, as was discussed in Chapter 3. Recognition of any benefits related to increased tourism as a result of the MPA could also be promoted.

Around half of the case studies focused on promoting awareness of such benefits, using approaches and techniques such as those discussed above in the 'raising awareness' section. This could be argued to represent a form of positive feedback, recognition of these benefits being a potentially important means of promoting support for conservation measures and cooperation with them. The potential of such positive feedback as providing a vital link in the co-evolution of social and ecological resilience is discussed further in the next chapter.

Knowledge incentives

Introduction: good enough knowledge

Uncertainty was previously discussed (Chapter 3) as a major challenge for the governance of MPAs, particularly in relation to establishing cause–effect links between certain activities and observed or potential impacts on conservation features. Knowledge incentives essentially aim to address governance challenges related to uncertainty, accepting that decisions will always have to be taken under a degree of uncertainty. Incentives in this category therefore aim to promote the use of combined sources of different knowledge in order to better inform governance decisions, recognising when the knowledge is good enough to justify a decision. Local users are recognised as having a fine-scale knowledge of the ecosystems they use, this being an important rationale for place-based management (Wilson, 2006), as was discussed in Chapter 4.

Promoting collective learning

The need to combine and integrate the traditional knowledge of local users with the scientific knowledge of 'experts' is increasingly recognised as essential for the adaptive management of social-ecological systems, local day-to-day knowledge and historical understandings of longer-term patterns being essential for the environmental feedback and governance response elements of adaptive management (Folke et al., 2005). Rather than attempting to combine the two types of knowledge *post hoc*, it is important to promote collective learning partnerships between local users and incoming 'experts'. These two streams of knowledge should be considered as complementary rather than competitive or conflicting, locals having a fine-scale knowledge built over years and even generations of living and working in a given place, scientists having wider-scale knowledge that is often related to meta-theories, enabling comparisons across different areas to further develop theories on the basis of applications and learnings in different contexts. It is therefore particularly important to promote mutual respect (based on social capital) amongst local people and scientists for the validity of each other's knowledge, as this is arguably an essential basis for collective learning through various forms of knowledge partnerships.

There were elements of collective learning in most if not all of the case studies, to varying degrees, several of the case studies providing particularly notable examples (Box 10).

Box 10 Case study examples of collective learning approaches

GBRMP

The consultation with stakeholders provided an opportunity for their knowledge on the distribution of different activities and conservation features to inform and influence the design of the zones. The connections that traditional aboriginal owners have with the marine environment and their cooperative management practices have been collated into the 'Story Place' database as a resource for researchers and managers.

California

Joint fact finding workshops were organised between stakeholders and scientists, which both improved the information base on which MPA proposals were based and improved the acceptability of the proposals. Users were also able to provide their knowledge of the features and uses of marine areas and explore the consequences of different network design options through interactive GIS systems. The MPAs are also being monitored on a partnership basis between stakeholders, state agencies, research scientists and NGOs.

Seaflower

Joint planning workshops between users and scientists helped build trust which in turn promoted the sharing of knowledge as a basis for MPA design.

Galápagos

Fish stock monitoring undertaken with the participation of fishermen.

Tubbataha

Standardised resource monitoring protocols agreed and applied by local people in collaboration with scientists in order to integrate the principles of science with traditional knowledge and to guide the process of learning.

Os Miñarzos

Academic scientists, fishermen and NGO researchers collaborated in the surveys and monitoring studies that informed the design of the MPA proposal, providing for the integration of scientific information with traditional ecological knowledge. GIS system developed to collate geographical, environmental and fishery data, including operations and catch data provided by fishermen.

Isla Natividad

The fishing cooperative initially began experiments to close and monitor areas and these were further developed in collaboration with NGO and academic scientists, guided very significantly by the fishermen's knowledge.

Agreeing approaches for addressing uncertainty

This incentive is focused on addressing the need to explicitly recognise the challenges raised by scientific uncertainty and agree approaches to address such challenges, e.g. agreeing rules for the interpretation and application of the precautionary principle, decision-making under uncertainty, and adaptation in the light of emerging knowledge. There is a great deal more emphasis in the social-ecological resilience literature on adaptation in the light of emergent knowledge generated through collective learning, e.g. Folke et al. (2005), the precautionary principle perhaps being regarded as too much of a legal and expert scientific doctrine for a concept based on self-governance. This principle is, however, a key rationale for no-take MPAs (Lauck et al., 1998; Murray et al., 1999) and given the major challenges that uncertainty raises for MPA governance, it will be considered here as a potentially important approach for addressing uncertainty. Furthermore, it could be argued that the case study MPAs that are no-take or include NTZs are applying the precautionary principle, but several case studies provide more specific illustrations of this (Box 11).

Box 11 Case study examples of agreed approaches for addressing uncertainty

GBRMP

The precautionary principle is required to be adopted in decision-making by the GBRMP Act and adaptive management in the light of monitoring feedback is central to policy and practice. One representation of the precautionary principle is that a given type of activity in a given zone is, by default, prohibited unless it is specifically provided for in the zoning regulations.

Darwin Mounds

The application of the precautionary principle in environmental decisions is a legal requirement under the treaty on which the EU is based. This was important in gaining an emergency regulation to restrict demersal trawling to protect the reefs under the EC's Common Fisheries Policy (CFP) before the MPA designation was adopted.

California

The legal basis of the MPAs specified that their design be based on the 'best available science', which helped avoid debates on how much evidence was required and minimised related delays.

Galápagos

The management plan stipulates that the precautionary principle must be taken into account in decision-making, but in reality many decisions are compromised either by a lack of precaution when they are being taken or a lack of implementation, one of the grounds for the latter being uncertainty.

Tubbataha

The precautionary principle was employed as the basis for the decision to designate a 10 nautical mile partially protected buffer zone around the no-take MPA.

Os Miñarzos

It was acknowledged at the outset of the MPA initiative that it was very difficult to predict the fisheries and ecological benefits of protection therefore the initiative should be considered as a pilot experiment.

Independent advice and arbitration

This incentive is focused on agreeing platforms for independent advice and/or arbitration in the face of conflicting information, uncertainty and views on the role of the precautionary principle, as a means of implementing approaches such as those discussed above. For such platforms to be effective, the validity of the independent advisor(s) and/or arbitrator(s) needs to be accepted, so that where agreement cannot be reached amongst users and other actors involved in the MPA, those on the platform have the remit to reach a decision or advise on the issue and the role of the precautionary principle, enabling deliberations and decisions to proceed on this basis, rather than governance processes becoming deadlocked by disagreements on a complex issue to which there is no definitive answer. Several case studies provided illustrations of the use of such a platform (Box 12).

Box 12 Case study examples of platforms for independent advice and arbitration

Wash

One of the roles of the public inquiry concerning this MPA was to resolve conflicting claims as to how the precautionary principle should be interpreted and applied. In other issues, independent experts were requested to resolve the question of the proportion of shellfish stocks that was required to provide food for migratory bird populations, and to carry out an assessment of the state of shellfish stocks on a mudflat.

California

A science advisory team consisting of a range of government and academic scientists deliberated and agreed on the guidance provided to the stakeholder group, subject to some further adjudication and interpretation by the Blue Ribbon Task Force.

Cres-Lošinj

There was a dispute between NGO scientists, backed by peer reviewed reports, and an academic scientist over the significance of the potential impacts of a proposed marina on dolphin populations. Supporters of the marina promoted the latter's views, the role of an independent arbitration panel to resolve this dispute being identified by the NGO as a critically important need.

Legal incentives

Introduction: the role of legal incentives

The role of the state and of laws set and enforced by the state are key to the co-evolutionary hierarchical governance concept discussed in Chapter 4, particularly in the final section on social-ecological resilience. It was discussed that whilst legal institutions may confer a degree of stability, their rigidity is considered by many neo-institutionalists to inhibit adaptability and thereby resilience. This argument will be returned to in the next chapter as it is central to this book, but it is important to recognise in the context of these discussions on the role of legal incentives that whilst there may sometimes be a need for a degree of legal flexibility in order to provide for adaptive management and local discretionary action, inflexible legal incentives are sometimes the only means of providing for conservation objectives to be achieved.

Hierarchical obligations

This incentive is focused on legal obligations to achieve particular objectives from higher institutional levels (international-regional/federal-national/state) that require effective MPA conservation, including the potential for top-down interventions if these obligations are not met. Such obligations, that set the direction and targets through specific objectives, but not necessarily the means of meeting them, can provide an important source of steer towards effective MPA governance. The majority of the case studies were subject to such targets, including all those in countries that have ratified the CBD or are party to the WSSD (Table 1). Those case studies that were managed primarily by the government under a clear legal framework (Approach I, Table 15) were understandably significantly steered by hierarchical obligations from a regional-national level and this steer was very influential in the governance processes. Those case studies classified as managed under the other management approach categories were understandably steered by a combination of obligations and other factors. Box 13 outlines some examples of the role of hierarchical obligations in various case studies.

Box 13 Case study examples of the role of hierarchical obligations

GBRMP

The federal GBRMP Act is primary legislation that has very considerable legal and political influence, this being one of the key factors behind the relative effectiveness of the MPA. Furthermore, the designation under the World Heritage Convention (outlined in Chapter 1) is having a considerable influence through interventions of the World Heritage Committee, including the related 'Fight for the Reef' NGO campaign (fightforthereef.org.au), as a political deterrent to port expansion and other development proposals that could lead to the GBRMP being officially categorised as 'in danger'.

Darwin Mounds, NE Kent and Wash

These MPAs are designated as Special Areas of Conservation (SACs) under the EC Habitats Directive, this being a supranational regional obligation. All EU states are obliged to implement all such directives as a legal requirement under the treaty on which the EU is based and this has provided a very strong legal incentive for the effective governance of these MPAs, e.g. the implementation of a 'revised approach' to better protect inshore SACs, following a threat of referral to the European Court of Justice by NGOs (Jones, 2012).

California

The obligation and related mandate to designate a network of no-take MPAs for specific conservation objectives under the MLPA was essential to the process, as were the Act's requirements for a combination of top-down and bottom-up approaches.

Sanya

Originally only subject to a municipal ordinance in the absence of a higher legal framework, but now subject to the state Regulation on Nature Reserves, under the authority of the provincial government, though there is a need for a stronger steer from the central state as some conservation obligations are still being undermined by an alliance of corporate tourism and municipal government development priorities.

Seaflower

Designated under a specific national law for this MPA which aims to implement various regional and international policy priorities, such as the target for MPAs under the CBD and the UNEP convention on the protection of the marine environment, but there is arguably a lack of clear national strategic steer, this being the only nationally declared protected area that is managed locally by a regional autonomous environmental authority (CORALINA). This has proved effective, as indicated by the success of CORALINA's legal actions to prevent the state from going ahead with proposals to license oil exploration and production activities in the MPA on the grounds that this would breach CBD obligations.

Galápagos

Designated under a specific national law for this MPA but there is a lack of clear national strategic steer through obligations for MPAs in general. The disproportionate political influence of the Galápagos electorate is a factor that has led to some MPA objectives being undermined by other political priorities, but interventions from an international level under the World Heritage Convention have been very significant in prioritising MPA objectives over other national and municipal political priorities.

Baleia Franca, Pirajubaé and Cres-Lošinj

The lack of binding obligations to effectively achieve the objectives of these MPAs is arguably a key reason why they have proved to be ineffective in the face of major driving forces. These forces came from the state itself in relation to the highway construction project that drastically undermined the effectiveness of the Pirajubaé MPA. In the case of the two Brazilian MPAs, one way of improving their effectiveness may be the extension out to sea of the politically encouraging and enabling role of the CBD and related institutions that have improved the effectiveness of terrestrial MPAs in Brazil. In the case of Cres-Lošinj, Croatia's plans to become a member of the EU could be a key way of improving effectiveness, as accession requires that all directives be fulfilled, including the species conservation obligations of the Habitats Directive, this MPA being focused on the conservation of bottlenose dolphins and cetaceans being a priority species under the directive. The NGO Blue World has played a key role in this respect in its involvement in notifying the EC of the ineffectiveness of this MPA under the Habitats Directive.

Capacity for enforcement

This incentive is focused on ensuring that sufficient state capacity, political will, surveillance technologies and financial resources are available to ensure the equitable and effective enforcement of all restrictions on all local and incoming users, including addressing the driving forces of incompatible trends in exploitation activities. This is the critically important 'sharp end' of governance, as the potential for cooperation amongst those willing to adhere to MPA restrictions is inevitably undermined if a minority are able to breach restrictions for their own personal and short-term gain. Whilst neo-institutionalists may place a great deal of faith in community or peer sanctions, they also recognise that some capacity for enforcement by the state to penalise determined local or incoming users, who stand to gain considerable economic benefit by breaching restrictions, is required for effective governance, as was discussed in Chapter 4.

This recognition appears to be strengthening and spreading amongst those involved in governing MPAs. Locally managed marine areas (LMMAs – www. lmmanetwork.org) have been pursued and promoted with considerable enthusiasm in recent years, these being the logical extension and application of neo-institutional governance theories, with their focus on the benefits and merits of self-governance by local communities in specifically defined places, particularly in LEDCs. A recent conference on enforcing MPAs, organised by the NGO WildAid in San Francisco in November 2012, included a session on 'MPA experiences from the field' which included many LMMA case studies. A common feature of these case studies was a focus on the bottom-up origins and processes of such designations in these relatively remote coastal and island tropical marine areas. However, another common and critical feature of these case studies was the need for top-down enforcement to complement local approaches, as there was insufficient enforcement capacity at a local level to ensure protection, particularly from incoming fishermen, who tend to aggressively resist detection or detention. This highlights how most, if not all, MPAs are dependent on a sufficient enforcement capacity to counter driving forces, capacity that only the state can realistically offer, even if there is a great deal of support from participating local users, e.g. a voluntary warden cannot be expected to challenge and detain incoming fishermen who are breaching MPA restrictions but who are likely to use violence or other retributions. There have also been cases of violence against official and voluntary wardens, including gun, machete, dynamite and acid attacks, which have led to serious injury and even death. In the face of such major challenges, arguably only sufficient top-down enforcement capacity by the state can be effective.

The question of how bottom-up approaches can be combined with the need for top-down enforcement is central to this book and is one that will be considered in the next chapter. For now, it is worth highlighting some examples of enforcement capacity issues from the case studies (Box 14).

Box 14 Case study examples of enforcement capacity issues

GBRMP

Emphasis on an integrated approach to enforcement across a range of federal and state agencies, e.g. GBRMP Authority, fisheries agency, police, coastguard, customs, etc., with formal agreements on shared responsibilities and inter-linkages. VMS is obligatory on all fishing vessels in the MPA but enforcement in some more remote areas remains a challenge.

Darwin Mounds

Enforcement occurs as part of the Scottish government's capacity to enforce the CFP, with mandatory use of VMS and electronic log books, coupled with air and sea patrols. Pelagic trawling is permitted so any vessel travelling at less than five knots is investigated, as this speed indicates demersal trawling.

Wash

Enforcement is undertaken primarily by the regional inshore fisheries authority, the obligation to do so recently having been strengthened through a 'revised approach' from the central government agency. iVMS, using smart phone technology rather than satellite GPS, is planned to be introduced to promote the compliance of vessels.

US NMSs

Enforcement relies largely on the capacity amongst different sectoral agencies and the lack of resources for a dedicated enforcement capacity is a particular challenge.

California

Enforcement is primarily undertaken by the California Department of Fish and Game but they have a relatively small number of vessels and wardens, recently reduced by public funding cuts, so their capacity is being augmented by other agencies, local governments and citizen groups.

Sanya

Restrictions on fishing and the activities of tourists are strictly enforced by wardens employed by the MPA authority (funded by tourism companies) along with guards employed by hotels. Seaflower

The lack of enforcement of regulations is a key problem, particularly fishing in NTZs and breaching access and technical restrictions in partially protected zones. This is partly due to the focus of the existing enforcement capacity on addressing drugs, arms and people trafficking. Peer enforcement approaches are being developed to improve compliance by local vessels, including funding from the GEF for patrol vessels, wardens, etc., and the allocation of more national enforcement capacity to improve compliance by incoming vessels is being promoted.

Galápagos

VMS has helped exclude illegal incoming vessels and enforce restrictions on local vessels, and an Automatic Identification System (AIS) has recently been introduced to improve this, but ensuring that these are fitted and operational on all local fleet vessels will be a challenge. IUU fishing by local and incoming fishermen remains a challenge due to a lack of capacity to intercept and detain vessels.

Karimunjawa

Enforcement is hampered by insufficient financial resources for patrols and an underlying lack of political will for enforcement, in the face of strong forces driving infringements, particularly market demand coupled with local poverty, but enforcement is improving as more infringements are being reported by users to community rangers funded by the government and NGOs, and by the improved integration of the MPA and district fishery authority's enforcement programmes.

Wakatobi

Similar challenges to above with additional challenge of poor links with Bajau communities. Enforcement has recently been improved by funding from NGOs and tourism operators for fishery patrols.

Tubbataha

Permanent presence of 10–12 MPA authority, navy, coastguard and municipal personnel on a ranger station is effective through observations and patrols, but maintaining sufficient capacity is very dependent on NGO funding.

Os Miñarzos

Surveillance is undertaken mainly by a private company contracted by the regional government and under the coordination of the MPA management body, but this has been reduced due to financial restrictions and this poses a major challenge as local fishermen lack the capacity to enforce restrictions.

Isla Natividad

The members of the fishery cooperative have the capacity to enforce no-take MPA restrictions on locals, but enforcement on incoming boats is a major challenge, as they are also often engaged in drug running, etc., so more state enforcement is needed, but this is lacking due to a focus of the enforcement capacity on other crimes, along with inefficiency and corruption.

Great South Bay

Enforcement capacity largely provided by municipal authorities but this is hampered by a lack of commitment from some authorities and confusions amongst town, county, state and federal enforcement agencies when violations are reported.

Chumbe

The private enterprise that created this MPA funds surveillance by unarmed rangers, in collaboration with local fishers. They rely primarily on education and warnings, but there is recourse to the arrest of persistent poachers by armed police based on the island.

Baleia Franca and Pirajubaé

Enforcement capacity falls far short of that required to ensure effectiveness, due to a lack of funding and an underlying lack of political will, e.g. enforcement of *all* environmental regulations in Baleia Franca (130km coastline, 1560km^2, including 10 cities) is undertaken by five employees with no boats, only two cars and an operational budget of less than $15,000.

Penalties for deterrence

The enforcement capacity to detect and detain illegal fishermen needs to be supplemented by the judicial capacity and the political will to apply it in order to impose sufficient penalties to provide a deterrent. This is generally not a particular challenge in the government-led MPAs (Approach I) as these countries have a relatively high state capacity and the judicial system is underpinned by federal and regional obligations. However, it is more of a challenge for the MPAs in the other categories as the countries in which these MPAs occur tend to have a relatively low state capacity, so the judicial systems tend to be relatively slow, there are often other priorities for the judicial systems to address and there may not be higher level obligations that can encourage the pursuance of prosecutions for breaching MPA restrictions. Many MPA case studies from the other governance approach categories illustrate how the successful prosecutions against those who breach MPA restrictions are undermined by a range of factors, such as long delays in proceeding with prosecutions, suspects being only briefly detained and then released after a non-binding warning has been issued, and prosecutions being withdrawn and amnesties being offered for political purposes, including as a result of corrupt links between the offenders and the state.

That is not to say that successful prosecutions have not been achieved in these case studies, but there is the tendency that such prosecutions tend to be relatively low in number compared with the number of infringements and that the penalties are often not sufficient to promote the future compliance of the offenders and other users who may otherwise have the potential to cooperate. Whilst the severity of penalties must take social justice and equity issues into account, it is also important to recognise that a lack of enforcement and prosecution capacity can critically undermine both the potential of some users to cooperate and the effective achievement of MPA objectives through the continuation of activities that impact on a given MPA. Effective MPAs are always dependent on a sufficient capacity to detect, detain and deter the minority of users who are inclined or driven to breach restrictions, as voluntary cooperation alone cannot realistically be relied upon.

Protection from incoming users

MPAs can confer major benefits for local users by providing for a degree of legal protection from incoming users who may not share the long-term stewardship goals that often motivate the involvement of local users in initiatives such as a MPA. This is particularly important given that exploitation by non-locals is widely recognised as representing one of the main threats to biodiversity, and potential alliances between the state and local users to fend off such threats is argued to be one of the principal benefits of co-management (Borrini-Feyerabend, 1999). Such protection is a growing priority given the growing influence of the driving forces discussed above, particularly the growing reach of fishing vessels and the related challenge of interloping fishermen who have no interest in local sustainability as they move from area to area in search of short-term opportunities. It is also an example of how ends–means synergies can be created between the objectives of conservationists and the objectives of local users (Murphree, 1994), in that conservationists regard local protectionism as a means of achieving conservation ends, whilst users regard local protectionism as an end to be achieved through the means of conservation.

   This incentive is closely related to the economic incentive of assigning property rights to local users, as is discussed in a previous section, but such rights require legal reinforcement, as local users rarely have sufficient enforcement capacity to exclude incoming users, as was discussed above. Several of the case studies provide examples of how protection from incoming users served as an incentive in governance processes (Box 15).

Box 15 Case study examples of protection from incoming users

Sanya

Access to fishing zones in MPA is restricted to local small-scale fishermen, excluding incoming industrial trawlers.

Galápagos

Licences to fish and rights to operate tourism vessels are restricted to legal Galápagos residents. Further protection is provided by restrictions on immigration and the repatriation of illegal residents.

Os Miñarzos

TURFS restricted to fishermen with a track record above a minimum number of days, in order to favour local vessels, exploitation by incoming fishermen, including recreational divers, being a key motive for the community-based designation of this MPA.

Isla Natividad

Only cooperative members can harvest fishery resources, which excludes incoming fishermen.

Attaching conditions to property rights

This incentive is focused on ensuring that the assignation of property rights for some activities to particular users does not undermine the achievement of conservation objectives and fulfilment of related obligations. This is achieved by attaching legally binding environmental performance conditions to such rights, breaches of which can lead to the withdrawal of the rights and/or prosecution. Concerns about assigning property rights were previously discussed under economic incentives, in relation to which Symes (2000) concludes that the state must act as the regulating authority and that responsibility and powers of sanction to conserve marine ecosystems must ultimately reside with the state. This supports arguments that the state can and should attach conditions to such property rights, so that state steer towards achieving wider conservation objectives can be applied indirectly through such property rights 'in the shadow of hierarchy' (Chapter 4), i.e. allocating property rights should be a means of decentralising authority, rather than relinquishing it. This is consistent with Hannesson's (2005) assessment that a combination of state control and rights-based approaches appears to be preferable. Several case studies provide examples of how legal conditions have been attached to MPA property rights (Box 16).

> # Box 16 Case study examples of attaching conditions to property rights
>
> ## Sanya
>
> Tourism user rights had several conditions attached, particularly that the developments comply with the MPA restrictions, and whilst it is debatable whether this condition has been met, the state could require compliance or withdraw the rights when they come up for periodic review.
>
> ## Galápagos
>
> The rights to operate tourism vessels ('cupos') include the condition that they cannot be rented to foreign-owned cruise vessels, but enforcement of this condition could be more effective.
>
> ## Os Miñarzos
>
> The submission of annual stock assessments is a condition of the TURF in order to ensure consistency with sustainable fisheries policies.
>
> ## Isla Natividad
>
> Renewal of fishing concession is contingent on annual stock assessments which demonstrate that the stocks are being managed sustainably.

Cross-jurisdictional coordination

This incentive is focused on ensuring that the decisions taken under policy frameworks for other sectoral activities, such as fishing, coastal development and terrestrial run-off, are consistent with and contribute to the achievement of MPA objectives. Other government authorities besides that which has responsibility for MPAs are generally responsible for such sectoral activities, which can have very significant impacts and can thereby undermine the achievement of MPA objectives. Legal requirements for cross-sectoral integration can therefore be very important, along with the political will to ensure such integration, accepting that an approach which relies solely on voluntary cross-sectoral cooperation is unlikely to be effective without some steer and related obligations from a higher institutional level. Cross-sectoral integration is an important element of implementing the ecosystem approach and/or ecosystem-based management, given that human and ecological inter-connections across jurisdictional boundaries and land–sea divides often need to be addressed under these approaches. Several case studies provide examples of strengths or weaknesses in this respect (Box 17).

Box 17 Case study examples of cross-jurisdictional coordination issues

GBRMP

The GBRMP Act takes precedence where other sectoral legislation and policies at a federal or state level are inconsistent with the achievement of the GBRMP's objectives. This is augmented by a formal Intergovernmental Agreement, which further promotes the integration of other sectoral policies, e.g. recent measures through the Reef Water Quality Protection Plan initiative to reduce land-based agricultural run-off as a means of addressing eutrophication issues in the GBRMP. However, closer coordination with fisheries management and coastal development policies is needed.

Wash

Measures to catalyse the implementation of fishing byelaws to protect biogenic reefs in the MPA are, at the time of writing, in the process of being introduced, 16 years after designation. These have been catalysed by the government's 'revised approach' policy which requires such integration, even though this integration was, strictly speaking, legally required under the 1994 regulations that implement the Habitats Directive.

Darwin Mounds

Effectiveness was promoted by the implementation under the CFP of a ban on demersal trawling in the MPA, though this is one of only two such permanent restrictions specifically to protect MPAs throughout the EU marine area that is governed under the CFP. It remains to be seen whether the reformed CFP will improve the prospects for the integration of fisheries policies with MPAs (Qiu and Jones, 2013).

US NMSs

From a legal perspective the US NMSs and National Marine Fisheries legislation have equal status so the latter should be consistent with the objectives of the former, but it can be a political challenge to gain fisheries restrictions to support the effectiveness of NMSs.

Sanya

Integration of tourism development with the MPA plan remains a challenge, but conditions attached to 'sea user rights' (see above) could address this.

Galápagos

A lack of coordination between the GNPS, the navy, the air force and municipal governments has hampered enforcement but this is improving, with the encouragement of the World Heritage Committee and as Ecuador steadily recovers from a long period of political instability and chronic underfunding.

Clear and consistent legal definitions

This incentive is focused on ensuring that there is clarity and consistency in legally defining the objectives of the MPA as well as general and zonal use restrictions, jurisdictional boundaries, responsibilities of different authorities, etc. Such clarity and consistency is very important in promoting the implementation of other legal incentives, particularly the cross-jurisdictional coordination discussed above. It is also important in relation to several other incentives in other categories, as the implementation of many other incentives will be undermined if the legal definitions of the MPA's objectives, restrictions, boundaries, etc. are unclear or inconsistent. Many of the MPA case studies illustrated the benefits of having such clarity and consistency, the legal and policy framework for the GBRMP perhaps being a model in this respect, though this was an incentive that it was recognised could be further improved.

Clarity concerning jurisdictional limitations

This incentive is focused on promoting clarity and openness concerning the jurisdictional limitations of the MPA legislation, i.e. recognising what driving forces, activities and impacts cannot be directly addressed by the MPA legislative framework and exploring means of addressing such factors. Whilst the incentive above was focused on being clear about what activities and related driving forces the legal and policy framework for the MPA *can* address, this incentive is focused on being clear about what activities and related driving forces *cannot* be addressed. This is important in preventing misunderstandings and frustrations amongst users and regulators if it is perceived that certain impacts beyond the jurisdictional limits of the MPA are being ignored. It is arguably even more important in enabling discussions and initiatives to address the impacts and the driving forces that are outside the jurisdictional limits of the MPAs, in that the legal and policy framework does not directly provide for them to be addressed, but which can be addressed through other political and legal routes.

One of the case studies illustrates the importance of this incentive both in identifying which driving forces are undermining the effectiveness of the MPA but which cannot realistically be addressed by the incentives within its jurisdictional limitations, and in motivating initiatives to address these driving forces through other means in wider jurisdictional and political frameworks.

The GBRMP's outlook reports have identified the impacts of climate change as having the potential to very significantly impact the MPA's coral reefs, through a combination of ocean acidification, ocean warming and sea level rise. However, it is also made clear that mitigating the effects of climate change is beyond the jurisdiction of the GBRMP authority, though it is critical that these impacts are reduced if the MPA's coral reefs are to survive. This has highlighted the threats posed by climate change to these iconic reefs at a national and international level, this being one of many factors that *should* encourage national and international initiatives to reduce anthropogenic greenhouse gas emissions. Whilst it is debatable how effective this has been in actually contributing to the effectiveness of climate change mitigation initiatives, it does nonetheless illustrate the importance of recognising the impacts of wider-scale climate change on this MPA and promoting initiatives to reduce these impacts through measures than can only be taken forward through wider-scale political and policy frameworks.

Legal adjudication platforms

This incentive is focused on employing legal adjudication and other formal and widely respected decision-making platforms to address and regulate conflicts which cannot be resolved amongst the users and regulators within the MPA's governance framework, through binding judgements. The judgements or decisions of such adjudication platforms need to be legally recognised as binding, subject to appropriate checks and appeals procedures (see below). This is particularly important where there is a proposal for a new activity or the modification of an existing activity which could lead to significant impacts on a given MPA and could thereby undermine its effectiveness in achieving its objectives, or where existing or proposed restrictions are considered or are claimed to have social, economic or cultural impacts that raise social justice and equity issues. These platforms are often provided with independent scientific advice, therefore their decisions will often incorporate an element of the 'independent advice and arbitration' knowledge incentive discussed previously. Several of the case studies provide examples of the benefits of such independent adjudication platforms (Box 18).

Box 18 Case study examples of role of legal adjudication platforms

Wash

A proposal by shellfish ranchers to deploy sonic scaring devices to deter seabirds that were feeding on the stock led to a public inquiry, whereby an independent inspector is appointed to take representations from all parties and make a recommendation to the government, which then reaches a binding decision. The recommendation to reject the application did create some tensions but it enabled governance processes to move on (Roberts and Jones, 2009).

Darwin Mounds, NE Kent and Wash

The European Court of Justice (ECJ) has legal authority over all MPAs in the EU designated under the Habitats Directive and its findings are binding on all national governments. Whilst no specific case related to these MPAs was actually referred to the ECJ, the mere potential for referral led to several significant governance decisions in these MPAs, such as that above and the 'revised approach' related to several incentives discussed in this chapter for the Wash.

Seaflower

A government decision to grant oil exploration and development licences was challenged through a 'People's Action' by the MPA agency CORALINA and referred to an independent legal tribunal, which supported the challenge and referred it to the High Court. This referral, along with an official opinion from Columbia's 'National Controller' that the licences were in breach of the CBD, led to the licences being revoked.

Transparency, justice and fairness

This incentive is focused on establishing legal provisions to ensure transparency and fairness in MPA management processes, e.g. statutory requirements for public access to information, provision for appeals by users and regulators, and conducting public hearings. Where legal incentives are being employed, it is essential that the process includes legal checks to ensure that the MPA legal and policy process is being adhered to, and that other laws and policies concerning social justice are also being adhered to, recognising that various direct and indirect users of a MPA could have recourse to such procedures, as could MPA regulators.

Such appeals could be on procedural grounds, evidence/knowledge grounds, social justice grounds and/or environmental justice grounds. Such legal provisions for transparency, justice and fairness are important if a societally acceptable balance between promoting effectiveness in achieving MPA objectives and promoting equitable outcomes is to be achieved. However, it can be very challenging to tease effectiveness and equity concerns apart, hence the need for judgements. This is why it is important to have independent legal adjudication platforms (as discussed above), and legal requirements for open and fair approaches to addressing equity and justice issues. Most if not all of the case studies included provisions for transparency, justice and fairness, and the examples above of the role of legal adjudication platforms also serve as examples of how these provisions are implemented. Some concerns related to justice issues in the case studies will be discussed in the next chapter in the context of the cross-cutting issue of equity.

Participative incentives

Introduction

These incentives are focused on providing for the participation of local users in decision-making and other governance processes. The focus of neo-institutionalists on self-governance was critiqued in Chapter 4, but this was with regards to their resistance to state steer. Neo-institutional arguments are premised on the view that attempts by the state to impose objectives and control decision-making and other governance processes will critically undermine the capacity for self-governance by local users, therefore the state's role should be confined to facilitating self-governance and assisting with the enforcement of decisions taken by local people. Whilst these arguments are critiqued, this is certainly not a rejection of the importance of participation. The co-evolutionary hierarchical governance concept recognises the important of all three sources of steer – state, market and civil society – hence the focus of these discussions on five categories of incentives, including participative incentives. There is a great deal of excellent guidance available on how to provide for stakeholder participation in natural resource governance processes, particularly through co-management, e.g. Borrini-Feyerabend et al. (2007) and this section will not attempt to represent such guidance. Instead, it will aim to briefly introduce each participative incentive in the context of the MPAG empirical framework, including some particularly significant examples of their interpretation and application in the case studies. However, given that all the case studies involve different combinations of various participative incentives, summaries of their specific application in relation to each case study will not be provided in this section. Instead, there will be a final discussion on the general trends in employing participative incentives amongst MPAs in the different governance approach categories (Table 13).

Rules for participation

This incentive involves establishing clear rules on how the participation of different interest and user groups will be provided for in governance processes on collaborative platforms and the role that such participation will have in decision-making and other governance processes. This raises a number of inter-related challenges. Whilst participation by consultation theoretically gives all potentially affected local people the opportunity to be represented, this is widely considered to provide too minimal a level of influence on deliberations and decisions to represent co-management. However, if a greater level of influence is provided for by involving local users in committees or workshops, it is important to recognise that it is not feasible to involve a large number of people in deliberations and decisions through such collaborative platforms. This generally means that particular groups of users have to be represented by selected individuals on such platforms. This, however, raises further challenges related to several questions, such as how to delineate particular groups of users, given considerations such as the geographic scale of representation, i.e. particular groups of users that operate in different areas may request separate representatives, but how do you decide the spatial scale across which representation will be provided for? Another consideration here is the delineation of particular groups of users, as small-scale fishermen, for instance, may argue that each group that uses a particular gear type should be represented, but how do you decide the level of disaggregation that groups of users should be afforded?

These are just some of the challenges of deciding how to provide for participation. There then follows another set of challenges related to the actual participation process. What degree of influence can users expect to have, whether by consultation, representation or other means, given the potential for different degrees of decentralisation? On what basis will decisions be reached, given that different representatives will have different vested interests, amongst which there may be internal (between different uses) and basic (between utilitarian and conservationist values) conflicts, therefore consensus on significant issues is often unattainable? Given the challenges that representatives face as 'boundary spanners' (Tushman, 1977), i.e. spanning the boundaries between the internal deliberations of the collaborative platform and the actual views and interests of the constituents they are trying to represent, how can processes be designed to minimise these challenges and potential tensions, e.g. when decisions to which a representative is party on a collaborative platform are subsequently challenged by their constituents?

These are but a few illustrations of the many challenges and questions raised when providing for the participation of users on collaborative platforms. This incentive is aimed at establishing rules that help address such challenges and questions, preferably on a proactive basis, but also accepting that such rules may often also have to be established on a reactive basis. Of course, this raises the question of who decides on such rules, e.g. whether this is on a more top-down basis, given that the state often decides whether and how users will participate in protected area governance (Borrini-Feyerabend, 1999), or whether it is itself on a deliberative, participative process, perhaps drawing on parallel cases for guidance. In the interests of avoiding complicated and potentially circular arguments, this question is left open.

Establishing collaborative platforms

This incentive is focused on designing and using governance structures and processes that provide for collaborative deliberations through the participation of the users of a given MPA, as well as the regulators, e.g. user/regulator committees, planning workshops, participative GIS, postal consultations on proposals that provide for detailed feedback, etc. Some guidance attempts to rank such approaches along a continuum, depending on the degree to which power or authority is delegated or transferred to local users, the rationale being 'the more, the better', but given the focus of this book of combining top-down and bottom-up approaches through the co-evolutionary hierarchical governance concept, this would not be appropriate. This incentive therefore basically involves establishing structures and processes that are appropriate to the context of a given MPA and that provide for an appropriate degree of user participation in deliberations and decisions on MPA planning and management issues.

Neutral facilitation

The operationalisation of the collaborative platforms incentive will often involve workshops and committees in which various sectoral and regulatory representatives will participate in deliberations to discuss and agree the development and implementation of management measures for a given MPA. Whilst these can be facilitated by one of the actors, this can lead to bias in the way the deliberations are steered, or perceptions of such bias. Facilitation also involves many specific personality traits and skills for which professional facilitators are specifically selected and trained. Whilst engaging such professional independent facilitators may have significant financial implications, the benefits gained in terms of progressing deliberations and reaching decisions through their expertise and experience that maximise the potential for agreement and cooperation amongst MPA actors will often more than outweigh the costs. This incentive is therefore focused on recognition of the advantages of such independent, neutral facilitation and of the importance of building the related costs into the initial planning of the MPA processes and structures.

Independent arbitration panels

As with the knowledge and legal incentives, processes involving participative incentives may lead to an impasse in negotiations on which agreement cannot be reached, or on which the view of a relatively independent panel of people could help progress governance processes. The panel should consist of people who are respected as having relevant expertise and knowledge but no actual stake in the particular MPA. It can be very challenging to organise and gain the resources for such a panel and often even more challenging for them to agree on recommendations that the actors involved in the MPA will respect and agree to. These challenges and the relatively recent recognition of the potential benefits of such an independent arbitration panel are amongst the reasons why this is perhaps one of the least applied participative incentives, so the sole example of this is worth discussing.

The process to design a network of MPAs under the MLPA is the only case study that would appear to have attempted to employ this incentive. The Memorandum of Understanding related to the MLPA specified that four Blue Ribbon Task Forces (BRTFs) be formed, one for each sub-region of California's coastline. Each BRTF consisted of eight people drawn from various public and private sector organisations, who were appointed by California's Secretary for Resources. Working in collaboration with the science advisory team, the BRTFs reviewed the marine reserve network proposals developed by the Regional Stakeholder Groups, which often included alternative configurations, and made final MPA network configuration recommendations to the Fish and Game Commission. This had the final decision-making authority and was not necessarily bound by the BRTF recommendations, but the final decisions often closely followed these recommendations. Whilst some stakeholders questioned the impartiality of the BRTF members and objected to the final MPA network configuration, the BRTF did play a significant role in making recommendations where agreement could not be reached amongst stakeholders, and also played a role in the initial implementation of the selected MPAs.

Whilst this example raises some issues, such panels appear to have considerable potential to progress governance processes in a way that is deemed as fair and impartial as feasible, by making recommendations on issues emerging from stakeholder deliberations on which agreement cannot be reached. They arguably represent a good alternative or supplement to legal adjudication panels, particularly where they draw on independent scientific advice, as discussed in the section on knowledge incentives. This incentive is focused on exploring the potential to employ such a panel in a given MPA initiative.

Decentralising responsibilities

This incentive involves allocating some roles and responsibilities to local users through a clear management structure, whilst maintaining an appropriate degree of steer by higher level state organisations in order to ensure that strategic conservation objectives are effectively achieved and that any related obligations, such as those discussed above in the legal incentives section, are fulfilled. Managing expectations in this respect can be particularly important by being realistic about the degree of autonomy and influence that local users can expect, given that whilst the state may have allocated some responsibilities, it should not completely relinquish all powers to local people if the state's responsibilities to achieve wider-scale societal objectives and higher level legal obligations are to be fulfilled. This was previously discussed in terms of three levels of decentralisation, in relation to the need to address scale challenges in Chapter 5, which can also be considered in terms of the degree to which responsibilities are decentralised to local people:

- Deconcentrated: transfer of responsibilities for implementing decisions but not taking decisions, these being retained by higher level state organisations
- Delegated: transfer of some decision-making responsibilities with a degree of control from the central government over key aspects of policy, particularly setting objectives and targets that decisions taken under delegated responsibilities must fulfill
- Devolved: the transfer of maximum feasible, but not necessarily total, decision-making responsibilities.

The replacement of the word 'power' used in the discussions on decentralisation in Chapter 5 with 'responsibilities' here is no coincidence, given that it would be inappropriate for the state to relinquish power and the argument in Chapter 4 that the concept of sharing power is too slippery and complicated.

Peer enforcement

In general, the more top-down an MPA's governance approach is, the greater the need for state enforcement capacity discussed above in the context of legal incentives. One of the many benefits of more bottom-up approaches to MPA governance is their potential to promote both the cooperation of individual users with MPA measures, due to their respect for and involvement in the process by which these measures were agreed, and to promote the potential for users to promote each other's compliance, widely referred to as 'peer enforcement'. This is also less widely referred to as 'graduated sanctions from other users' and 'internal enforcement' in the design principles for CPRs (Ostrom, 1990, 94–101), though these conditions also recognise that such sanctions are often combined with sanctions from officials (discussed above in terms of enforcement capacity), in what is described as 'quasi-voluntary compliance', as it is based on both peer pressure and the potential for recourse to legal enforcement sanctions. Focusing on peer enforcement, the detection of transgressors can be by voluntary wardens, other users or other community members, whilst sanctions can involve various approaches, such as the social marginalisation of transgressors, loss of social status, loss of access to community resources and even threats of vandalism and violence. By way of a transition towards legal enforcement capacity, peer enforcement can also involve reporting transgressors to the appropriate regulatory state authority.

There are also more subtle forms of peer enforcement, whereby the knowledge that most if not all other users are complying with a norm creates an assumption that such compliance is normal (hence the term 'norm') and expected. There are, as is discussed above in the section on enforcement capacity, limits to what such peer enforcement can achieve, particularly with regards to incoming users and determined rogue locals, referred to as freeriders in neo-institutional parlance. However, it is also important to recognise that, given the logistical challenges of policing large marine areas, in reality, much of the day-to-day enforcement of MPA restrictions must rely on peer enforcement. This incentive is therefore focused on providing for peer enforcement by maximising its potential, through the use of other participative incentives, as well as by other measures such as resources for community wardens and responding to reports of infringements.

Building social capital

As was discussed in Chapters 4 and 5, the concept of social capital is central to neo-institutional theories on environmental governance. The previous critiques in this book of these theories were focused on their arguments that social capital amongst local users in a given place will provide for self-governance, and that this should be the basis for natural resource governance, without any interference through control or steer by the state. Social capital has been discussed in terms of the degree to which actors (users, regulators, etc.) reach and implement decisions together through their professional and social networks, based on the development of mutual trust amongst actors, and on confidence that where an actor is willing to cooperate in the implementation of collective decisions, their cooperation will be reciprocated by other actors through collective actions (Jones and Burgess, 2005 drawing on Ostrom, 1990, 1998, 1999).

Studies have found that social capital is generated particularly well through face-to-face meetings/discussions (Ostrom, 1998), as well as through related factors such as transparency and equity. Furthermore, it has been found that social capital is generated rather than being consumed through its use, and that this can lead to an 'upward spiral' (Ostrom, 1998) of cooperation and confidence that cooperation will be reciprocated amongst actors, whilst erosion of trust through lack of cooperation, transparency, equity, enforcement, etc. can lead to a 'downward spiral'. For instance, users who bear some short-term costs as a result of their cooperation with collective decisions to achieve strategic objectives are likely to decide to stop cooperating if other 'freeriding' local or incoming users are gaining short-term benefits by not cooperating, e.g. fishing in no-take MPAs. This incentive is focused on the recognition that generating and building on social capital through the promotion of social activities such as face-to-face meetings, social gatherings, celebrations of achievements, beach cleans, etc. amongst MPA actors, as well as other approaches related to many of the other incentives, is as important in the concept of co-evolutionary hierarchical governance on which this book focuses as it is in the neo-institutional concept of place-based self-governance that it critiques.

Bracing linkages

Bracing social capital is 'a kind of social scaffolding' that aims to 'strengthen links across and between scales and sectors but only operates within a limited set of actors'. It is developed through strategic, selective, focused and instrumental initiatives, and its development is often guided by state actors in order to reinforce partnerships with local users in such a way that strategic objectives and obligations can be fulfilled (Rydin and Holman, 2004). Rydin (2006) discusses how bracing social capital is a concept that considers the horizontal and vertical linkages between the state and civil society that go beyond the locality and are necessary to achieve strategic policy objectives that are increasingly prevalent in natural resource governance contexts. The concept thereby relates to the need for institutional linkages, discussed in Chapter 4, in order to address the scale challenges posed by human and ecological interconnections, as well as to the importance of links based on trust discussed above. This incentive is focused on the strategic development of such bracing linkages, recognising that these will often need to be instrumentally developed by state actors involved in the MPAs in order to gain the participation, support and cooperation of key people within user and regulatory communities in the local and wider context, in the hope that these key people will then be able to influence other people in their networks and promote wider support for participation in and cooperation with various other MPA incentives. The development of bracing social capital can, as its name implies, strengthen and improve the resilience of a governance framework.

Building on local customs

One of the risks of imposition discussed in Chapter 4 is that MPAs pursued mainly on a top-down basis, i.e. mainly employing legal incentives, can over-ride or even 'squash' customary ways of sustainably utilising natural resources through traditional practices, norms and customs amongst local people, as they become marginalised by unholy alliances between NGO, state and corporate actors, for which MPAs can be vehicles. In order to minimise this risk, this incentive aims to promote consistency with and respect for local traditions, customs, norms and practices in the development of the MPA governance framework through the development of 'hybrid' institutions (Cinner and Aswani, 2007) that combine traditional institutions with MPA incentives. This can be challenging, however, as whilst this incentive contributes to the objective of maintaining traditional uses (Chapter 2), it was previously highlighted that such uses can be affected by driving forces and drift towards more intensive commercial exploitation, potentially undermining MPA effectiveness. This incentive therefore aims to maximise the integration of customary management practices into the development of MPA governance incentive frameworks, provided they are compatible with and contribute towards the fulfilment of marine biodiversity conservation objectives and obligations, recognising that compromises on both sides may need to be negotiated in such 'hybrid' institutions (Cinner and Aswani, 2007).

Its operationalisation will usually involve a baseline assessment of existing traditional uses and the customary practices associated with them, followed by an assessment of their compatibility with the achievement of the conservation objectives and of how incentives to promote the effectiveness of the MPA can build on and be combined with customary practices. It is important to recognise the potential need for compromise in modifying customary practices to make them more compatible with the MPA *and* for revising the MPA's objectives and incentives to accommodate customary practices.

Potential to influence higher institutional levels

A defining feature of the co-evolutionary hierarchical governance concept discussed in Chapter 4 is that the process of negotiated compliance can lead to users in a given MPA having the capacity to influence higher level institutions through vertical linkages which provide for feedback that leads to revisions at higher institutional levels, i.e. a two-way process. This is an important difference between the top-down 'command-and-control' government concept and the co-evolutionary hierarchical governance concept, which combines top-down and bottom-up approaches. This incentive is focused on designing the governance framework in such a way as to provide for influences on higher level institutions, often through various other participative incentives, ensuring that actors involved in a given MPA are aware of this potential and engage in upward feedback initiatives, e.g. consultation responses, feedback questionnaires and collaborative committees, and ensuring that higher level institutions respond to such feedback, whilst also ensuring that the structure of the governance framework still provides for effectiveness in fulfilling the MPA-related strategic societal objectives and obligations. Again, as with the above incentive, this is a process that generally involves compromises on both sides, as some higher level institutions may not be negotiable given the need to achieve societal objectives and related legal obligations, but there is usually scope for some negotiation and compromise.

The Seaflower case study provides an illustration of how this incentive can operate in a manner that increases effectiveness. This decentralised MPA was largely driven by local users working alongside CORALINA, which operates as a regional environmental authority on an autonomous basis. When the national government granted oil exploration and development licences, many local people involved in the MPA objected and CORALINA instigated a 'People's Action' which successfully challenged the decision and led to the withdrawal of the licences. This represents a very significant influence at higher institutional levels, albeit an influence that employed the leverage of hierarchical obligations under the CBD and the fulcrum of an adjudication platform, as is discussed above in the context of legal incentives. However, this case still demonstrates the influence via these mechanisms that local users involved in a decentralised MPA had on central decisions that were related to nationally important economic interests, but that could have adversely affected local users. The potential for such influence could serve as a significant incentive for local users to participate, whilst accepting the potential for compromises between achieving local utilitarian priorities and wider-scale conservation objectives.

Overview of participative incentives

As participative incentives, like interpretative incentives, are so widely recognised and applied, details of how they have been applied in specific case studies have not been discussed, apart from the California and Seaflower examples. It is, however, worth briefly overviewing trends in the application of the participative incentives across the five governance approach categories, including a few case study examples.

The six case studies assigned to the government-led category (Approach I, Table 15) do apply some of the participative incentives, but it is important to note that, with the exception of California, all of these case studies essentially confine user participation to feedback through consultations, with the state having most if not all of the decision-making authority. This partly reflects the tendency that the state has the capacity to directly implement MPAs, as the relatively high state capacity metrics indicate. Whilst other participative incentives are applied to provide for the input of different users, the governance of these MPAs is steered by the state, as the title of this category implies, this partly also being a reflection of their roots in national legislation and sometimes higher level obligations, e.g. to the EC for the UK case studies and, to a lesser but still significant degree, to the World Heritage Committee for the GBRMP. Users were given a greater role in the California case study, but the need for these MPAs was driven by state legislation and the final decisions about them were all taken by the state. The application of the participative incentives does, however, still provide for the views of users to be aired, considered and taken into account in all six of the case studies assigned to this governance approach category.

The seven case studies assigned to the decentralised category (Approach II) predictably and appropriately apply the participative incentives in a manner that gives users a considerable influence on governance processes and decisions, partly because the state lacks the capacity to implement MPAs on a direct basis, as the relatively low state capacity metrics indicate. By definition, the incentive of decentralising responsibilities has been very influential in these cases, but there is considerable variation in the way that the participative incentives have been applied and the influence that local users have on decisions amongst these case studies. In Sanya, most responsibilities have been decentralised to the local municipal government, which is working very closely with the tourism development sector. In this case participative incentives have been lightly applied, if at all, and the local users have actually had very little influence, some having lost both their land and their access to marine areas in a manner that has raised equity and social justice concerns. The Ha Long Bay case study represents more of a municipal planning initiative that affects some users, but mainly in positive ways, with decisions mainly being taken by planning authorities, with little provision for the significant application of participative incentives. In the other five case studies (Seaflower,

Galápagos, Karimunjawa, Wakatobi and Tubbataha) there has been very significant application of participative incentives, with local users having a considerable influence over decisions, working alongside state and NGO actors, the latter often both supporting the decentralised governance processes and having a considerable influence, as is discussed in the next chapter.

The two case studies categorised as being managed primarily by local communities (Approach III: Os Miñarzos and Isla Natividad) appropriately rely largely on steer through participative incentives, with user representatives on collaborative platforms having a major role in steering governance processes and reaching decisions. Many of the other participative incentives have also been applied in these case studies. However, in the case of Os Miñarzos, local users lacked the capacity to enforce the MPA restrictions on incoming users, and this has led to a greater role for legal incentives which could undermine the influence of participative incentives. In the case of Isla Natividad, the fisheries cooperative members themselves had the capacity for peer enforcement on local users, though there are some equity concerns in this respect as the majority of locals are not cooperative members and have thereby lost their customary access to shellfish resources. However, the fishery cooperative again lacks the resources to enforce restrictions on some incoming users and is seeking additional state enforcement capacity in this respect, though this seems to have less potential to undermine the influence of participatory incentives given that the cooperative has formal community property rights.

The two 'private sector' MPAs (Approach IV: Great South Bay and Chumbe) arguably had very little provision for participative incentives, as they are managed by a NGO and a private company respectively. However, in the case of Great South Bay, neighbouring local government representatives participate in collaborative planning and it is hoped that their state enforcement capacity can be better applied to help enforce the restrictions. In the case of Chumbe, local fishing village representatives participate in the MPA's advisory committee, but the private company is not bound to adopt their recommendations, though there have been no significant disagreements to date. The failure of the three case studies categorised as ineffective (Approach V: Baleia Franca, Pirajubaé and Cres-Lošinj) is arguably not due to any lack of participative incentives, as these case studies have provided for several such incentives to be applied, giving local users significant roles. However, the lack of legal incentives and political will, coupled with the influence of significant driving forces through major infrastructure and economic development projects, have undermined if not competently countered the influence of participative steer.

Summary

This chapter has been a long and detailed analysis of the different ways in which different incentives have been applied in different case studies. This length and detail is indicative of the focus of the MPAG framework on deconstructing governance issues. This is important as a means of: (a) 'teasing apart' the 36 incentives in order to better understand their individual governance roles; (b) providing for a more detailed understanding of the incentives, potentially serving as a 'menu' for those involved or interested in the application of such incentives in governing other MPAs; and (c) 'bringing the incentives to life' by illustrating how they have (or have not) been successfully applied in different contexts. The next chapter will focus on bringing together or synthesising these findings, including discussions of some cross-cutting themes and general conclusions.

Chapter 8

Resilience through diversity

Introduction

This final chapter will begin with an overview of the incentives that were most frequently cited as being used and needed in the case studies, followed by discussions on the role of leadership, the role of NGOs and issues related to equity in relation to the case study findings. Where the previous chapter focused on the role of individual incentives, this chapter will then focus on how diverse incentives interacted to promote resilience in the social-ecological systems in the case studies, and on the vital role of the state and of legal incentives in reinforcing the incentive frameworks against the perturbing effects of forces that can drive the systems towards ineffectiveness and unsustainability. The overall conclusions of this study will then be discussed, focusing on its implications for governance theories, for research and for practice in relation to MPAs and wider natural resource issues, followed by a closing synopsis of the findings of this study.

Incentives used and needed

The empirical framework for the 20 case studies not only asked the participants to identify and discuss the incentives that had been used in their case study, but also to discuss which incentives are particularly needed in order to improve the governance framework by better addressing conflicts and thereby potentially increasing the MPA's effectiveness. These findings were analysed by totalling the total number of times each incentive was cited as being used and the total number of times each was cited as being needed across all the case studies. The incentive most frequently cited as being used was the interpretative incentive *raising awareness*, which all the case study MPAs understandably sought to do through various different means. The next three incentives most frequently cited as being used were each from a different category: *provision of NGO and private sector funding* (economic), *promoting collective learning* (knowledge) and *establishing collaborative platforms* (participative), each of which was used in 15 of the case studies.

By contrast, the four incentives most frequently cited as being needed to improve the governance framework were all legal incentives: *capacity for enforcement* (cited as being needed in 14 case studies), *cross-jurisdictional coordination* (10 case studies), *clear and consistent legal definitions* (six case studies) and *hierarchical obligations* (five case studies). This indicates that whilst the incentives that are most frequently identified as being used are from a variety of categories, the incentives that are most frequently identified as being needed tend to be legal incentives. The implications of this preliminary finding are discussed later in this chapter, but it is worth noting here that it would seem that legal incentives are most often considered to be particularly needed in order to improve MPA governance frameworks, even in MPAs that adopt a decentralised (Approach II), community-led (Approach III) or private (Approach IV) governance approach.

The broad patterns in the use of different categories of incentives amongst the case studies in different governance approach categories were briefly discussed in Chapter 6. Suffice it to say here that a diversity of incentives from across the range of the different incentive categories was used by the MPAs across the different governance approach categories, with legal incentives predictably being particularly frequently applied in government-led case studies, and economic incentives being particularly frequently applied in decentralised, community-led and private case studies. It is also worth briefly discussing the incentives that were discussed as being particularly needed to improve effectiveness amongst case studies in each of the governance approach categories.

Government-led MPAs (Approach I) were considered to be particularly in need of further *cross-jurisdictional coordination* mechanisms, this being a challenge in countries with a relatively high state capacity, but also a relatively complicated legal and policy framework, with different sectoral jurisdictions. This indicates that cross-jurisdictional integration mechanisms to ensure that other sectoral policies contribute to MPA effectiveness, rather than undermine it, often need such legal coordination mechanisms. Furthermore, given the general need for some steer in this respect from higher state levels, such coordination also requires the political will to use such mechanisms. The GBRMP would seem to have addressed this need relatively well, with legal requirements for other sectoral policies such as coastal development and land-based pollution to contribute to rather than undermine MPA effectiveness, though some challenges remain. The US NMSs, California and Wash examples illustrate that there is often significant potential for improved cross-jurisdictional coordination in government-led case studies.

Decentralised MPAs (Approach II) were considered to be most in need of legal incentives. They were particularly lacking in *capacity for enforcement*, coupled with a lack of political will to utilise the available capacity, often manifested as a lack of funding and human resources for surveillance to detect infringements, and a lack of sufficient *penalties for deterrence* for detected infringers. This is perhaps because providing for economic development and sources of subsistence represents a more important political priority. Most MPAs in this category are facing multiple and strong driving forces, including growing coastal populations, increasing domestic and international demand for seafood, and the rapid development of tourism, most of which cannot be fully controlled and mitigated through actions at the local level, requiring interventions at national or even international levels. Whilst decentralised governance approaches are very appropriate for countries with a relatively low state capacity, there still needs to be a sufficient enforcement capacity and the political will to apply it to not only detect those who breach MPA restrictions, but also to apply sufficient penalties to deter those who have been prosecuted and other people who could potentially do so. The influence of *hierarchical obligations* appeared to have the potential to encourage national governments to implement such legal incentives, e.g. Seaflower, Galápagos.

Decentralised MPAs were also considered to have the potential to benefit from strengthening legal incentives for ensuring *transparency and fairness* in the governance processes and in the equitable sharing of costs and benefits derived from the MPA. Economic incentives cited as being needed for improving MPA governance in the decentralised MPA case studies include the *assignment of property rights* to communities and traditional users (Sanya) and more focus on *promoting alternative livelihoods* and sustainable business enterprises owned by local communities (Karimunjawa). Finally, the strengthening of knowledge incentives for *improving collective learning* (Sanya, Wakatobi) and *agreeing approaches for addressing uncertainty* (Galápagos) is also identified as being needed to improve governance in some MPAs in this category.

Community-led MPAs (Approach III) were previously discussed as being particularly vulnerable to changes in the wider socio-economic and political environment (Chapter 6), therefore they may benefit from the strengthening of legal incentives related to *hierarchical obligations* and *cross-jurisdictional coordination* in order to address these vulnerabilities. Both the Os Miñarzos and Isla Natividad case studies could also be considered to need improved *capacity for enforcement* and *protection from incoming users* in order to further reinforce these MPAs against the driving forces that could perturb their governance frameworks and undermine their effectiveness. The Isla Natividad case study also raised particular equity concerns related to the loss of access to fisheries by some local people. The incentives of *attaching conditions to property rights*, *promoting alternative livelihoods* and *transparency and fairness* could help address these concerns, as is discussed later in this chapter. These examples illustrate that reinforcement against driving forces and incoming users, along with addressing local equity concerns, are arguably likely to be issues that need particularly close attention in community-led MPAs.

Private MPAs (Approach IV) were considered to have similar issues and incentive needs to community-led MPAs, in that they are vulnerable to changes in the political and economic environment, which may, for example, affect the land lease and decentralised management agreements entrusted to the private company or NGO, this being a particular concern in the Chumbe case study. As with community-led MPAs, private MPAs could be considered to be in need of the legal incentives of *hierarchical obligations* and *cross-jurisdictional coordination* in order to address these vulnerabilities. Political will and support from government authorities, in terms of providing and enforcing a legal and policy framework for conservation, is as important in private MPAs as it is in MPAs from other governance approach categories.

Ineffective MPAs (Approach V) were predictably in need of many incentives from most if not all of the categories. Whilst efforts have been made to more effectively govern these MPAs through the use of various participative incentives, these have largely been in vain and there is a particular need for legal incentives to address the strong driving forces that are leading to the effectiveness of these MPAs being undermined, coupled with a stronger political will to both develop such incentives and apply them. In the case of Cres-Lošinj, it seems that Croatia's plans to become a member of the EU offer most potential in this respect, as the government will be subject to *hierarchical obligations* under the EC Habitats Directive to protect, amongst other marine features, cetacean populations. Similarly, in the cases of Baleia Franc and Pirajubaé, similar but less binding *hierarchical obligations* under the CBD appear to have some potential to address the lack of political will to effectively conserve these MPAs. The particular need for economic incentives was also identified in these case studies in order to financially support and justify these MPAs.

Cross-cutting issues

The MPAG empirical framework described in Chapter 5 included three cross-cutting issues that were briefly reviewed. These will be returned to in this section, drawing on the findings of the case studies.

Role of leadership

There are doubtless many examples of important leadership roles that various people have played in the case study MPAs, too many to discuss in detail here. One particularly notable example is the lead role of Sibylle Riedmiller in establishing and effectively governing the Chumbe MPA in the face of political and economic challenges that most individuals and organisations would have considered insurmountable. Without her commitment, experience, passion and determination in leading this innovative initiative, Chumbe would not have been designated, nor would it continue to be the world's leading example of just how effective and equitable private MPAs can actually be.

One overall trend with regards to the role of leadership is that people can adopt such roles in relation to MPAs across the five governance approach categories and in support of initiatives across all five incentive categories. Certainly, the role of leadership is not confined to the role of promoting the participation and priorities of local users, e.g. particular people in government positions or in positions to lobby governments can take an important leadership role in developing the political will that is often critically important for the effectiveness of MPAs, as will be discussed later in this chapter. The participative incentive of *bracing linkages* is significant in this respect, as the establishment of links based on social capital between leaders at one or more government levels and between local leaders who are respected by their community or user groups can be important in reinforcing collaborative partnerships to designate and effectively manage MPAs.

It is also important to remember that, in the same way that there are dis-incentives for effective MPAs, e.g. short-term economic development opportunities related to marine resource extraction opportunities, discussed in the context of MPAG empirical framework as driving forces that MPA incentives must withstand, there are often people who can adopt strong leadership roles in objecting to, delaying and even derailing the progress of a given MPA or network of MPAs and thereby undermining effectiveness. Strong leadership in support of the MPA and the political will to take MPAs forward in the face of such counter-leadership roles is thereby of even greater importance.

Role of NGOs

NGOs come in a variety of sizes and play a variety of roles in different MPA contexts, from small NGOs focused on promoting the rights and welfare of local communities, to big international NGOs (BINGOs) focused on lobbying for more MPA designations and more effective approaches to governing MPAs at national and international levels, e.g. the campaign for very large ecosystem-scale no-take MPAs (GOL, 2011) discussed in Chapter 2. The findings of the case studies illustrate how NGOs can play a variety of roles, largely depending on which governance approach the MPA represents.

In the government-led MPAs (Approach I) NGOs can still play major roles. For instance, in the case of California, the RLFF was very influential in lobbying for the introduction of the legislation (MLPA) which made the MPA networks obligatory in the first place. The RLFF then went on to broker the public–private partnership through which the network design processes were largely funded by the RLFF but undertaken through collaborations between stakeholders, regulators and scientists. On the whole, though, NGOs tend to play more indirect roles in government-led MPAs, focused on providing additional funding to support MPAs, as well as on lobbying the state to designate more MPAs and more effectively govern them. For example, the Great Barrier Reef Foundation provides additional funds to support research to promote the resilience of this MPA, whilst various NGOs are involved in the 'Fight for the Reef' campaign which is lobbying against infrastructure developments, using the World Heritage designation and related potential for the listing of this MPA as 'in danger' as a legal and political lever. The threat by NGOs to refer the UK government to the EC for the insufficiently proactive and precautionary protection of MPAs designated under the EC Habitats Directive led to the introduction of the 'revised approach' which requires such protection. Greenpeace's legal challenge led to the extension of the UK's application of the Habitats Directive from territorial waters to the entire continental shelf, under which the Darwin Mounds were subsequently designated. These cases illustrate that NGOs can still have a very significant role in government-led MPAs.

In the decentralised MPAs (Approach II) NGOs tend to play more direct roles, in that they often provide a greater proportion of the funding and some governance responsibilities may be decentralised to them. They can also play a significant role in lobbying national governments to designate and more effectively govern MPAs. All of the MPAs in this governance approach category illustrate these roles, with the exception of Sanya, as NGOs currently have less of a role in governance processes in China. The Coral Triangle Initiative has been very influential in promoting a network approach to MPAs in the Philippine (Tubbataha) and Indonesian (Karimunjawa and Wakatobi) case studies, and three BINGOs play a key role in funding and supporting this initiative, alongside state and development bank partners. These and other NGOs also play a key role in the funding and operation of these MPAs, involving projects related to *promoting alternative livelihoods, promoting collective learning, raising awareness*, etc., alongside local users and various central, regional and local governmental bodies. Similarly, various large and small NGOs have played various roles in lobbying for the more effective governance of the Galápagos MPA, funding governance initiatives and leading on particular projects.

In the community-led MPAs (Approach III), NGOs have played key roles from the outset in supporting and working alongside local communities in designating and effectively governing the Os Miñarzos and Isla Natividad MPAs. The role of NGOs varies in the private MPAs. In the case of Great South Bay, a BINGO acquired the property rights and thereby adopted managerial roles, though they still need the support of municipal authorities in enforcing restrictions. In the case of Chumbe, a private individual financed and took the lead, NGOs playing only very minor supporting roles. NGOs also played key roles in the MPAs currently categorised as ineffective MPAs. The Cres-Lošinj MPA was initiated, funded and steered by a NGO, but the lack of political will and related legal incentives from the state led to its ineffectiveness. In the Brazilian cases, NGOs were influential in the designation of these MPAs. In all three cases, NGOs are playing important roles in lobbying, making legal challenges, engaging in collaborative planning with local users and other initiatives aimed at improving the effectiveness of these designations.

This overview of the role of NGOs indicates that they can play various important roles in governing MPAs across all governance approach categories. Certainly, the role of NGOs is not confined to promoting participation and self-governance. Whilst many of the concerns previously discussed in Chapter 5 may also apply, i.e.: (a) that NGOs neglect the diversity within 'the community'; (b) that they play a role in the tyranny of participation; (c) that there is a lack of evidence for their effectiveness in promoting participation and alleviating poverty; (d) that they impose conservationist values on local users; (e) that they can adopt a quasi-state role without the legitimacy and accountability of the state; and (f) that they can be a key actor in 'unholy alliances' that can displace and marginalise 'traditional' users, these concerns did not emerge as issues in these case studies. However, these case study analyses did not include the sort of in-depth ethnographic studies amongst actual users of the MPAs, which would be more likely to reveal such concerns, so it cannot be assumed that there are no concerns about the role of NGOs and their potential to capture the steer of a given MPA. The evidence from these case study analyses does indicate that NGOs are making significant contributions to the development of MPA governance frameworks that promote effectiveness through the roles discussed in this section.

Equity issues

It can be very challenging to balance effectiveness in achieving the conservation objectives for MPAs, often involving restrictions for particular users, against equity in the fair distribution of costs (related to use restrictions) and benefits (related to the achievement of objectives) arising from MPAs. As was discussed in Chapter 3, MPAs invariably generate basic conflicts between priorities and uses underpinned by utilitarian values, and those underpinned by conservationist values, through the very act of designating an area as 'protected', such conflicts being at the root of many governance challenges. These challenges were discussed in Chapter 4 in terms of achieving a balance or compromise between the risks of imposition and the risks of parochialism, recognising that there is a perverse symmetry in that certain sections of local communities can be marginalised and disadvantaged by both top-down and bottom-up approaches.[10] Whilst win-win solutions that address both priorities are discussed as the way forward for avoiding such basic conflicts, with conservationist wins on one hand through the achievement of MPA objectives by restricting activities, and utilitarian wins on the other through fisheries and/or tourism benefits by promoting the profitability and sustainability of such activities, the reality is that addressing basic conflicts often leads to costs for some people and benefits for others: 'winners and losers' rather than 'winners and winners'. MPAs often inherently influence the distribution of such costs and benefits and can thereby lead to criticisms that equity has been undermined in the pursuit of effectiveness.

Equity issues can also be considered in terms of social and environmental justice. Social justice relates to concerns that certain people bear an unfair proportion of the costs of MPA restrictions, including issues related to fishing as a 'way of life', e.g. subsistence or small-scale commercial fishermen being marginalised from governance processes and decisions and then being denied access to their usual fishing grounds through no-take or partial MPA restrictions. Environmental justice relates to concerns that certain people in a given area or larger groups of people in wider society bear an unfair proportion of environmental costs related to the impacts of particular uses of a MPA, e.g. the loss of the climate mitigation potential of certain marine habitats and of other ecosystem services as a result of the impacts of fishing. Equity thus appears to be a multi-faceted issue. As was discussed in Chapter 5, achieving a balance between social and environmental justice issues often involves trade-offs between effectiveness and equity, and balancing the need to provide for effectiveness and minimise inequities is often extremely challenging in MPA governance.

It is important to consider equity from the perspective of some particularly relevant incentives, as these provide a potentially important means of promoting equity and addressing related justice issues. Whilst many incentives are potentially relevant to equity and justice issues, some incentives are particularly important in this respect. The incentive discussed in the previous chapter of *legal adjudication platforms* provides three examples of how such platforms helped to address equity issues. Interestingly, whilst these examples will have included deliberations on social justice issues, the outcomes in these cases were mainly focused on addressing wider environmental justice issues by restricting certain activities in order to improve the effectiveness of the MPA in achieving its objectives. This is consistent with the tendency that MPAs tend to be mainly focused on addressing environmental justice issues and related basic conflicts, i.e. restricting certain uses (underpinned by utilitarian values) in order to fulfil MPA objectives (underpinned by conservationist values), thereby addressing wider societal concerns about declines in the health of marine ecosystems. However, it is also important to address social justice issues raised amongst local users in pursuing this environmental justice focus.

The use of adjudication platforms to address equity issues is closely related to the legal incentive of *transparency and fairness*, legal provisions for which are important to promote equity and justice, and compliance with which can be deliberated on by adjudication platforms. The legal incentive of *penalties for enforcement* also has potential equity and justice implications, as a balance needs to be struck between applying penalties that are sufficient to deter the prosecuted person and others, whilst not being unjustly harsh. The participative incentive of *building social capital* is also important in this respect as insufficient penalties can lead to downward spirals of cooperation and thereby the erosion of social capital, but unjustly harsh penalties that increase inequities can also potentially lead to such downward spirals. Governance approaches that promote equity and justice can contribute to the building of social capital, potentially leading to upward spirals. The economic incentive of *measures to reduce the leakage of benefits* is also particularly relevant to promoting equity, as it enables local people who may bear some opportunity costs as a result of MPA restrictions to gain a fair proportion of the benefits arising from the achievement of MPA objectives. The incentive of *providing compensation* potentially enables costs that cannot be offset by the benefits of effective MPAs to be offset by such payments.

It is also important to consider any equity and justice issues emerging from the case studies, given that the MPAG empirical framework explicitly sought such issues, and to consider the incentives that helped address them. In the government-led case studies (Approach I), such issues are addressed by ensuring that the legal and policy frameworks provide for transparency in addressing equity issues. This is achieved through the legal incentive of *transparency and fairness* by assessing the distribution of costs and benefits associated with a MPA and publishing the results, and of *legal adjudication platforms* by providing for recourse to appeal procedures, e.g. the legal adjudication platform employed in the Wash case study and the potential for such platforms in the Darwin Mounds and NE Kent case studies, including referral to the European Court of Human Rights. Whilst there are no doubt users related to the government-led MPAs who feel that they have suffered social injustices as a result of

MPA-related restrictions on their activities and incurring related costs, it is important to remember that such users do have legal rights of access to relevant information and of recourse to appeal processes and platforms, such rights being important elements of state capacity. There are also often a greater number of alternative means of earning a living through other economic opportunities in MEDCs and, as a last resort, there are relatively good social welfare provisions for those for whom social injustice related costs prove to be unmanageable, though these provisions are no substitute for promoting equity and access to justice.

In the GBRMP case, they also attempted to address social justice issues by applying the economic incentive of *providing compensation*, but the magnitude of these claims 'spiralled out of control' (Macintosh et al., 2010) and it is debatable whether this really represented an equitable outcome for society as a whole. The UK government, by contrast, has maintained a position whereby it will not pay compensation to fishermen who claim they are suffering a social injustice by bearing costs as a result of MPA restrictions. The factors behind the UK government's reluctance to pay such compensation include that: (a) marine fisheries represent a public right, therefore those exercising that right are not eligible for compensation from the public purse; (b) fishers can switch to alternative grounds to maintain their income, recognising that this can incur extra fuel/labour costs and raises not only extra risks related to vessels being forced to steam further but also risks related to effort displacement; (c) experiences with compensation claims arising from fisheries exclusions related to oil/gas and wind farm developments indicate that it can be extremely challenging to distinguish between fishermen who claim they rely on fishing in a given area in order to claim compensation, and those that actually do rely on such fishing, and, in the latter case, to determine what proportion of a fisher's income is derived from a given marine area (Jones, 2009).

This policy would appear to be wise in the light of the GBRMP experience and it seems likely that it will be adhered to in the UK unless fishermen can successfully claim that it breaches their human rights, which seems unlikely given that such rights do not necessarily extend to the right to earn a living in a specific way. Whilst UK fishermen do argue that they are suffering social injustices as a result of MPA restrictions and that they deserve compensation (Jones 2009), these arguments have not yet been referred to a court. It is notable that the US has adopted a similar position in relation to NMSs and the California MPAs, in that fishermen are not paid compensation for social injustices that they may claim to have suffered through bearing costs related to loss of access to fishing grounds as a result of MPA restrictions. Overall, though, the approach to addressing equity and social justice issues related to MPAs in government-led case studies in MEDCs with a relatively high state capacity would appear to be focused on the legal incentives of *transparency and fairness* and *legal adjudication platforms*.

Equity and justice issues arguably raise more major challenges in the case studies assigned to the other governance approach categories, as the majority of these are in LEDCs with relatively low state capacities, per capita incomes and human development indexes. These metrics indicate that there may be less access to information about the costs and benefits associated with MPAs, and less access to legal recourse, as well as fewer options for alternative livelihoods and/or social welfare provisions. LEDCs also tend to have relatively high proportions of their population living below the poverty line, indicating that more people are likely to be critically dependent on access to natural resources for their livelihoods. People who live on or near MPAs in LEDCs therefore tend to have a higher dependence on exploiting marine resources, so loss of access to such resources resulting from MPA restrictions can raise equity and justice issues that are a much greater concern than they are in MEDCs. Furthermore, people who have a high dependency on marine resources for their livelihoods but who lose access to these resources as a result of MPA restrictions may have few options other than to try and breach these restrictions in order to subsist, potentially both undermining effectiveness and exposing people to the risk of being penalised, thereby further worsening their situation and exacerbating inequities.

Some, if not all, of the case studies categorised as decentralised (Approach II) may include equity and justice issues, some particularly notable issues having been recognised in this study. In the Sanya case study, some local people claimed to have suffered injustices due to the loss of land and/or of access to marine resources as a result of the allocation of tourism development rights, coupled with MPA restrictions. These inequities have, to a degree, been offset by the economic incentives of *reinvesting MPA income in local communities* and *promoting alternative livelihoods*, and the legal incentive of *protectionism from incoming users*, but the economic incentive of *assigning of community property rights* to certain areas for traditional uses could help to better address them.

In the Wakatobi case study, the indigenous Bajau people have a traditional view of marine resources as being for subsistence use and they tend not to participate in the collaborative platforms for the MPA or take up alternative livelihoods. This has the combined effect of further alienating Bajau people if they are prosecuted for infringing NTZ restrictions in MPAs, or of undermining effectiveness through infringements that are not prosecuted, which tend to be the majority. The perceptions of these inequities are exacerbated by the key role that incoming NGOs and dive tourism operators are playing in this MPA. These inequities could be better addressed by the economic incentive of *assigning property rights* for traditional users, though this is challenging as the Bajau people may not recognise being confined to a specific zone given their nomadic lifestyle. Involving Bajau people in *raising awareness* initiatives and deliberations through *collaborative platforms* could help their understanding of the MPA and provide for their input to decisions and thereby their cooperation with them, minimising the risks of imposition, but they are resistant to official top-down initiatives such as MPAs as they are not part of their traditional way of life. Initiatives involving anthropologists in understanding and *building on their local customs* could also provide a way forward for reconciling effectiveness and equity issues, but this case illustrates the major challenges and dilemmas that equity issues can raise in MPA governance, as there is no incentive or combination of incentives that seems to readily provide a solution.

The community-based (Approach III) case study of Isla Natividad illustrates equity issues related to the risks of parochialism, as some local people who used to have customary rights of access to shellfish in the area but who are not fishing cooperative members have been denied access to these resources through the *assignation of property rights* to exploit shellfish stocks to cooperative members, which excludes non-members, though they may be able to gain employment in the processing and marketing of catches. In this case, what initially appears to be a flat and bottom-up governance structure actually seems to consist of a local 'hierarchy of wannabees', potentially raising equity issues related to the risks of parochialism (Chapter 4), though no such issues were specifically identified in this case. Were such issues to be raised by non-cooperative members, the top-down role of the state in providing for the rights of marginalised groups could become important, e.g. through the legal incentive of *attaching conditions to property rights* that require for some access to fishery resources by non-cooperative members in order to avoid the risks of parochialism. This case could also be considered to need improvements in the legal incentives of *transparency and fairness* and access to *legal adjudication platforms* to address such equity concerns, coupled with the economic incentive of *promoting alternative livelihoods*, such as local dive centres, and *reducing the leakage of benefits* and *protection from incoming users* in relation to maximising local employment in dive centres and reducing the activities of incoming dive boats.

No major equity issues were raised in relation to the private MPAs (Approach IV), though there may be fishermen who have been excluded from the fishery in the Great South Bay and Chumbe cases who may claim they have suffered a social injustice through a NGO or private company acquiring the property rights. Conversely, there may be local people who claim that these MPAs have addressed environmental justice issues by ending the impacts of fishing by closing these MPAs. Similarly, there may be local people who claim that the ineffectiveness of the case study MPAs in Brazil (Approach V) has raised environmental justice issues by allowing impacting activities to go ahead that have degraded their environment and damaged their traditional fishing activities. Similar concerns may be held by people who are concerned about the impacts of recreational boating on dolphin populations in the ineffective Cres-Lošinj MPA. These cases illustrate how more in-depth ethnographic studies amongst actual users of the MPAs may be needed to reveal some equity issues, as was discussed above in relation to the role of NGOs.

Whilst such research may reveal more equity issues and related social and environmental justice concerns, the findings of the case studies indicate that a range of incentives could help in addressing such issues and concerns. The legal incentive of *protection from incoming users* appears to have particular potential, especially given the related potential for ends–means synergies discussed in relation to this objective in Chapter 7, whereby compromises are reached between local users accepting some restrictions on their activities to achieve conservationist objectives, in return for protection against incoming users and related driving forces. Kothari's quote in a paper focused on 'protected areas and equity concerns' (Blaustein, 2007) concludes on an optimistic note in this respect:

My biggest hope is that we will get to a situation, across the world, where indigenous and local communities will demand the creation of protected areas, to protect themselves against outside destructive forces [i.e. incoming users and related driving forces], and to derive social and economic benefits from them without compromising on their conservation values.

The economic incentives of *reducing the leakage of benefits* and *promoting alternative livelihoods* also seem to have particular potential. However, the increasing faith in and emphasis on *promoting alternative livelihoods* as a win-win solution arguably needs to be tested by some more rigorous evaluations of the social, economic and environmental sustainability of such alternative livelihoods and the degree to which benefits gained from them really do offset losses incurred through loss of access to the resources related to MPA restrictions. The legal incentives of *legal adjudication platforms* and *transparency and fairness* also seem to have particular potential, but there is a need for a reasonably high state capacity to provide for these, sometimes including the capacity to address the thorny issue of corruption, involving inappropriate links between state and private sector actors.

Whilst particular incentives can be focused on as a means of addressing equity issues, it is important to consider that it is the interactions of a combination of incentives that arguably has the greatest potential to address such issues. Furthermore, many would argue that having effective MPAs is necessary to address wider environmental justice issues related to the impacts of damaging activities such as fishing and wider societal concerns about the health of marine ecosystems, and that these issues need to be balanced against local social justice issues. The nexus of effectiveness and equity issues remains a central and critical challenge for MPA governance, including balancing the risks of imposition against the risks of parochialism, and there are no simple solutions, the balance between effectiveness and equity and the acceptability of related trade-offs and compromises ultimately being a question for societal judgement.

Social-ecological resilience through incentive diversity

Having discussed the case study findings in relation to three cross-cutting issues, this section will return to the focus on the role of different incentives. Chapter 7 teased the role of incentives apart in order to gain an understanding of the role of individual incentives in the context of specific case studies. The previous discussions in this chapter focused on the need for particular incentives to improve the effectiveness and equity of MPA governance frameworks. This approach to considering the role of particular incentives is analogous to autecology, whereby the focus is on the relationship of an individual species with its environment. By contrast, this section will take an approach that is more analogous to synecology, whereby the focus is on the ways in which many species interact to constitute a structurally and functionally integrated, diverse and resilient ecosystem. This approach will consider the ways in which incentives interact to constitute a structurally and functionally integrated, diverse and resilient governance system.

An important trend from the previous discussions in this and the previous chapter on the role of particular incentives is that a given incentive often interacts with one or more other incentives in steering human behaviour towards the effective achievement of MPA objectives. For instance, several case studies indicate that the economic incentive of *assigning property rights* interacts with the legal incentives of *attaching conditions to property rights* and *protection from incoming users*, these being essential to the effective operation of property rights, whilst *transparency and fairness* and *legal adjudication platforms* are important to provide for accountability and equity. Other incentives could further support the role of property rights, such as *reducing the leakage of benefits*, *promoting collective learning, establishing collaborative platforms* and *peer enforcement*.

Other case studies indicate that the participative incentive of *decentralising responsibilities* needs to be combined with the legal incentives of *hierarchical obligations* and *clear and consistent legal definitions* in order to ensure that the decentralisation of responsibilities does not undermine the capacity to effectively achieve MPA objectives that represent wider societal priorities. Similarly, *capacity for enforcement* interacts with *penalties for deterrence*, both being necessary for *promoting profitable and sustainable fisheries*, which is itself further supported by *promoting recognition of the benefits, reducing the leakage of benefits* and *cross-jurisdictional coordination*.

The more you look at the interactions between different incentives and the more laterally you think, the greater the number of interactions. From this perspective, the many interactions between different incentives from different incentive groups can be considered as a web, the combination of the incentives and the interactions between them forming a governance system. This perspective led to an analysis of the interactions amongst the incentives in the 20 case studies, whereby we estimated the frequency with which one incentive supported, complemented and reinforced another. This analysis is represented in Figure 8.

Building on the discussions in Chapter 2 on the MPA objective of restoring marine ecosystems, one of the most significant developments in ecology in recent years is the recognition that having a diversity of different species and of different functional groups of species, with a complex web of strong and weak trophic interactions between them, confers ecosystem stability through functional redundancy, and adaptability through response diversity. In the light of the MPA case study analyses on which this book focuses, it is argued that the same rationale follows for MPA governance systems, i.e. having a diversity of different incentives and of different categories of incentives, with a complex web of strong and weak reinforcement interactions between them, provides for an effective and equitable MPA governance framework. This rationale applies to MPAs from all the governance approach categories, in that whilst the patterns of interactions may vary across governance approaches, the need for a diversity of incentives from a diversity of categories with many reinforcement interactions does not. From this perspective, ecosystems and governance systems appear similar when considered as social-ecological systems, as Figure 8 illustrates.

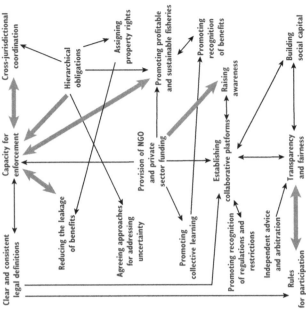

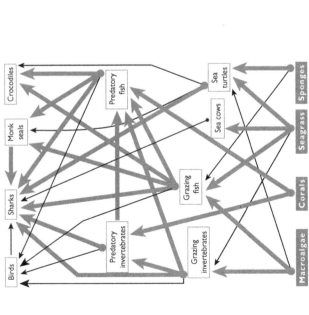

Figure 8 Trophic web and incentives web.

(Left) Simplified trophic web for typical tropical coral reef, including seagrass meadows, a thick arrow indicating a strong trophic interaction and a thin arrow indicating a weak one. Adapted from Jackson et al. (2001) with permission from AAAS. (Right) Incentives web – the interactions between different incentives, a one-way arrow indicating a one-way interaction with one incentive reinforcing another, a two-way arrow indicating a mutual reinforcement. Where an interaction was recorded in more than five case studies, it was considered to be a particularly important and strong interaction (thick arrow). Only 18 of the 36 incentives are represented in this figure, as some incentives were not cited as being used in these preliminary 20 case studies, some incentives have subsequently been added, or significant interactions with one or more other incentives could not be identified. This figure is for illustrative purposes and does not represent a definitive analysis. Adapted from Jones et al. (2013b) with permission from Elsevier; the supplementary material related to this paper includes a full description of the analysis on which this figure is based.

It soon becomes clear from this perspective that trying to understand a governance system by focusing on the role of an individual incentive, or even a particular category of incentives, is analogous to trying to understand an ecosystem by focusing on one species or one functional group of species, i.e. impossible. As was discussed in Chapter 5 (Table 11), economic incentives represent a market-based mode of governance, legal incentives represent a top-down mode of governance and participative incentives represent a bottom-up mode of governance, interpretative and knowledge incentives supporting all three modes. Given the need for a diversity of incentives from all categories, including reinforcement interactions between these incentives, the distinction between top-down, bottom-up and market-based approaches becomes blurred and less important, as incentives related to these approaches need to be combined to reinforce each other. Combining a diversity of appropriate incentives from different incentive categories in such a way that they reinforce each other is arguably what the statement that the 'design and management of MPAs must be both top-down and bottom-up' (Kelleher, 1999, xiii) actually means in practice, as diversity is the key to resilience in both ecosystems and governance systems.

In the light of this understanding, trying to work out which incentives replicate the functions of other incentives and/or appear to have no significant function and are therefore redundant is no more relevant than trying to work out which species in an ecosystem are redundant and therefore dispensable. Diversity is the key to resilience in both social and ecological systems as this provides for functional redundancy and response diversity. Therefore the concerns discussed in Chapter 4 that institutional redundancy may lead to fragmentation, duplication, inconsistencies and inefficiencies are arguably, themselves, redundant, given that incentives represent particular types of institution and the focus should be on maximising diversity rather than searching for redundant and thereby dispensable incentives. Whilst it is important to ensure that incentives complement and reinforce each other, rather than conflict or be inconsistent with each other, trying to avoid institutional redundancy undermines the quest for resilience through incentive diversity.

As was discussed in Chapter 4, it is important to recognise that whilst we may discuss the governance of social-ecological systems, the reality is that we only have the capacity to steer social systems, ecosystems being steered by ecological interactions and physical environmental conditions. Humans have the ubiquitous capacity to impact marine ecosystems and perturb their structure and function through, for example, trophic cascades induced by overfishing, but we do not have the capacity to actually steer them by purposefully intervening in ecological interactions and physical environmental conditions. Our influence on marine ecosystems is essentially focused on influencing human behaviour/activities through appropriate combinations of incentives in order to minimise our impacts. In the context of the governing of MPAs, this means restricting certain activities by combining different incentives, in order to reduce the impacts of these activities, as a means of conserving or restoring the structures and processes of marine ecosystems. This does not mean, however, that social and ecological systems are not coupled, given that humans have the capacity to negatively impact the structural and functional integrity and biological diversity of marine ecosystems, potentially reducing their resilience and the flow of ecosystem services they provide, i.e. to reduce the health of marine ecosystems, on which the health of humans depends.

Fortunately, we also have the capacity to positively impact the health of marine ecosystems, including through the designation and effective governance of MPAs, by applying appropriate combinations of incentives to restrict certain human activities and thereby effectively achieve MPA objectives. The resulting recovery of structural and functional integrity, biological diversity and resilience should also increase the flow of ecosystems services, through the recovery of fish stocks and thereby increases in CPUE, improved and sustainable tourism income, enhanced coastal defence value, etc. The realisation and recognition of these benefits should increase the potential of local actors involved in the governance of a given MPA to both cooperate with existing incentives and to propose and/or develop further incentives to further improve the health of the MPA and thereby its flow of ecosystems services. In this way, MPAs can be considered as vehicles to promote the effective governance of marine social-ecological systems through the following cause–effect linkages:

- the effective application of a combination of MPA governance incentives reduces the impacts of human activities;
- this increases the species diversity and thereby the health of marine ecosystems;
- this improves the flow of marine ecosystem services;
- this improves the sustainability and profitability of marine activities;
- these benefits increase the potential of local actors to both cooperate with existing incentives and to propose/develop further incentives in order to further improve effectiveness.

These cause–effect links between the effective governance of human activities through diverse incentives in governance systems and the improved health of our seas through diverse species in ecosystems have the potential to become cyclical. In this way, social-ecological systems have the potential for an upward co-evolutionary spiral of cooperation, effectiveness and resilience as a result of the direct steer of social systems through incentives, and the indirect steer of ecosystems through impact mitigation and the recovery of diversity. The 20 case studies on which this book is based provide evidence that a diversity of incentives from different categories reinforce each other through complex interactions, and that this promotes effectiveness in achieving MPA objectives, including the conservation and restoration of a diversity of species from different functional categories. There is some evidence that these increases, coupled with increases in the density and biomass of populations, have increased resilience (Box 19).

Box 19 Case study examples of effective MPA governance leading to increased resilience

GBRMP

One of the symptoms of the declining health of coral reefs in the GBRMP was an increased frequency of crown-of-thorns starfish outbreaks. In the face of uncertainty as to whether declines in fish populations and/or terrestrial run-off were exacerbating these outbreaks, the area of NTZs in the GBRMP was increased on a precautionary basis. Recent evidence indicates that recoveries in the health of NTZs have made them more resilient to such outbreaks, in that they suffer fewer outbreaks than comparable fished areas and thereby have a higher coral cover (McCook et al., 2010). This builds on previous evidence of an inverse relationship between levels of fishing intensity and the density of crown-of-thorns starfish on reefs in Fiji (Dulvy et al., 2004).

Isla Natividad

The relatively large body size and high egg production of abalone populations in the NTZs conferred resilience to the impacts of anoxia episodes related to ocean warming, enabling these populations to have an increased survival and recovery rate. Larval export from these populations contributed to the replenishment of populations that were smaller and less dense in fished areas and therefore had lower survival and recovery rates (Micheli et al., 2012).

Chumbe

Coral reefs in the no-take MPA were less impacted by coral bleaching related to ocean warming than surrounding unprotected reefs and they recovered sooner (Nordlund et al., 2013). The coral reefs in the MPA are recognised as being amongst the most resilient in the Western Indian Ocean to the impacts of climate change as a result of their high biodiversity levels (McClanahan et al., 2007; Maina et al., 2008).

Whilst additional evidence will enable the proposed cause–effect cycle and the related potential for an upward co-evolutionary spiral to be further explored, the findings reported in this book support the argument that a diversity of incentives from different categories provides for the conservation and restoration of the effectiveness of MPAs, and thereby of the health of the marine ecosystems they cover. Most if not all of the relatively effective MPAs are showing some evidence of the recovery of populations of species and of the diversity of species, this being a logical if not inevitable consequence of restricting activities that were impacting them, so it follows that this will be leading to the restoration of ecological resilience through increased species diversity. Furthermore, whilst MPA governance can only provide for the control of activities within MPA boundaries and of nearby and upstream activities, such as land-based pollution in the MPA's catchment, the case studies on which this book draws indicate that effective governance can conserve and restore the resilience of a MPA to wider-scale impacts, such as ocean warming related to climate change. Recognising that the three case studies in Box 19 represent government-led, decentralised and private MPAs, this argument potentially applies across MPA governance approach categories. This supports the general argument that the key to social-ecological resilience is incentive diversity, both of individual incentives and of incentives from different categories.

The vital reinforcing role of the state

In the context of the MPAG empirical framework, described in Chapter 5, effectiveness is the 'bottom line', and is evaluated by assessing the degree to which the impacts of local activities have been ameliorated through appropriate restrictions in order to provide for the achievement of the particular objectives of a given MPA. As was discussed in Chapter 6, all of the case studies are exposed to driving forces that could perturb the governance framework by promoting incompatible activities, thereby undermining effectiveness and sustainability. As was discussed in Chapter 7, various incentives from different categories are being applied amongst the MPAs from different governance approach categories in order to withstand the potentially perturbing influences of such driving forces and thereby to promote effectiveness.

A recurring theme in these discussions is that the state has a critical role in enabling and supporting the implementation of various incentives in MPAs from all the governance approach categories. Even in MPAs in categories II, III and IV (decentralised, community-led and private), governments still play a major role in providing a supportive legislative and policy environment and in controlling and mitigating the driving forces that cannot be effectively addressed at a local scale. The increasing diversity, reach and magnitude of driving forces means that in most MPAs, it is a matter of when, not if, such driving forces will eventually become strong enough to perturb or disrupt the MPA governance system and undermine effectiveness. Such potential has already been realised in several of the case studies, e.g. Sanya, Galápagos and Baleia Franca, while others were considered to be particularly vulnerable to such potential, e.g. Isla Natividad and Chumbe.

It appears that MPAs in countries that are emerging from a previous regime and rapidly engaging with the global economy, indicated by a relatively high GDP growth rate metric, e.g. China, Ecuador, Brazil, are particularly vulnerable, as the driving forces are relatively strong and there is a shortage of state capacity, resources and political will in these countries to address driving forces that promote economic growth. As the influences of globalisation and economic development priorities increasingly affect LEDCs, MPAs in these countries are likely to face similar challenges if resilient governance frameworks with a diversity of incentives, particularly reinforcing legal incentives, are not in place. It is argued that, despite the trend towards decentralisation in protected area governance, a degree of regulation and state control through legal incentives is necessary to withstand the potentially perturbing effects of driving forces, including those originating from the state. Therefore, it is appropriate that even the highest degree of decentralisation, referred to as devolution in Chapter 5, does not transfer total decision-making powers to lower level organisations, as this would amount to the state relinquishing power. It is necessary for the state to retain a degree of steer over decentralised governance approaches in order to address scale challenges and to fulfil wider-scale strategic conservation objectives and related obligations. The discussions in Chapter 7 and earlier in this chapter provide many examples of how legal incentives interact with and reinforce other incentives from other categories, these interactions also being illustrated in Figure 8.

An important trend in this respect is that legal incentives often provide for indirect steer or control, e.g. through *attaching conditions to property rights* that promote effectiveness and equity, as well as providing reinforcement for other incentives, e.g. *protection from incoming users* being vital for several economic and participative incentives. From this perspective, the need for legal incentives and state steer is not a return to 'command-and-control' or 'fortress conservation'. If used appropriately, legal incentives integrate with and reinforce many different incentives from different categories, thereby enhancing the resilience of the MPA governance system and providing important support for the implementation of incentives from all the other categories.

The metaphor of a building constructed of reinforced concrete springs to mind, with economic, interpretative, knowledge and participative incentives representing the concrete and legal incentives representing the steel reinforcement rods. In the same way as a building made of concrete but without reinforcement rods will soon buckle and fall, a MPA governance framework without a strong web of legal reinforcement incentives will soon be perturbed, eventually collapsing in the face of increasingly strong driving forces, leading to a paper MPA that is ineffective in achieving its objectives and in contributing to measures to address wider societal and scientific concerns about the declining health of marine ecosystems. Participative incentives make a very important contribution to the stability of the governance framework, along with economic, interpretative and knowledge incentives, but no matter how large the number or combination of these incentives, they cannot substitute the reinforcement role of legal incentives[19] in promoting not only the effectiveness of MPA governance frameworks in achieving conservationist objectives, but also their equity in not unfairly discriminating against marginalised people through local governance processes.

Political will is a particularly important element of the critical role of the state, as most legal incentives originate from and are implemented by the central state through a variety of democratic and judicial processes. The state also has the capacity to promote large-scale economic and infrastructure development projects that can undermine effectiveness, as the Pirajubaé case study (Box 13) vividly shows.

Political will at high government levels is also important for the legal incentive of *promoting cross-jurisdictional coordination*, and for minimising the risk of development proposals that contribute to the main government agenda of promoting economic development, whether such proposals arise from the private sector or the state itself. Building on the arguments that MPAs require legal incentives whether they are government-led, decentralised, community-led or private, it is clear that all MPAs require political will regardless of the governance approach being followed, the lack of political will being the main reason why three MPAs are categorised as ineffective in this study.

Furthermore, many economic, interpretative, knowledge and participative incentives rely on adequate state support, not only in relation to the need for reinforcing legal incentives, but also in relation to the need for financial resources and bureaucratic support. It is not only important for a sufficient state capacity to exist, there must also be the political will to apply this capacity to promote the effective governance of MPAs. Leadership by state representatives also often plays an extremely important role in progressing MPA designation and governance processes, preferably including from officials at high political levels. If a government lacks the political will to implement and effectively govern MPAs, it follows that the national MPA network will be inadequate in coverage and/or in effectiveness, often leading to a sparse network of paper MPAs, e.g. Brazil (Gerhardinger et al., 2011b).

Political will can be promoted in different ways, e.g. the GBRMP illustrates how the interpretative incentive of *raising awareness* was employed to generate public support for better conservation, this support then being used as a democratic mandate to promote political support for improved conservation measures; the Galápagos and Seaflower case studies illustrate how political will can be promoted by top-down *hierarchical obligations* related to the World Heritage Convention and the CBD respectively. Whether political will is generated from the bottom-up or from the top-down, or exists on a latent basis due to the commitment of one or more political leaders to marine conservation, there is little doubt that it is absolutely vital for the development of effective MPA governance frameworks.

Conclusions

These arguments and findings have implications that could be of interest to natural resource governance theorists, MPA governance researchers and MPA practitioners. These implications and the related conclusions will be briefly discussed leading to a general closing conclusion.

Implications for governance theories

These findings and the related arguments have significant implications for the governance analysis concepts discussed in Chapter 4. The focus of neo-institutional governance analysts on place-based self-governance arguably not only neglects the human and ecological interconnections between different marine places, but also neglects the need for top-down approaches and legal incentives, which the case study findings reported in this book indicate are vitally important elements of MPA governance frameworks. Due to a combination of their roots in alternative participatory development and adherence to Habermasian ideals, neo-institutional governance concepts remain resistant to the role of the state on the grounds that this interferes with self-governance by local actors in quasi-autonomous places. The role of the state is confined to facilitating and supporting deliberations amongst local people, and assisting in enforcing decisions, including through the provision and enforcement of sanctions, but not interfering with or undermining such deliberations, particularly through the imposition of conservationist objectives and related legal obligations.

Whilst these concepts recognise that self-governance institutions are nested or embedded in higher level and wider-scale institutions, which is particularly important given the scale of human and ecological interconnections, integration largely relies on faith in the potential of negotiations through horizontal and vertical linkages. Furthermore, the need to fulfil MPA obligations, as a contribution to meeting wider-scale targets in order to address growing societal concerns about the health of marine ecosystems, is essentially dismissed by some neo-institutional governance analysts, on the grounds that such obligations undermine and interfere with local self-governance regimes and traditional sustainable natural resource use practices, which are more effective than protected areas.

However, if the importance of legal incentives is accepted, including as a means of providing necessary sanctions against freeriders and protection from incoming users, it also has to be accepted that it is not realistic to expect the state to provide such incentives on a passive basis, with no involvement in or steer of local deliberations, particularly given that the state has responsibilities to address wider societal concerns and fulfil conservationist objectives and obligations related to MPAs. These responsibilities mean that whilst the state may decentralise some elements of governance to local actors, it would be irresponsible for the state not to maintain a degree of direct and indirect steer of local governance processes, i.e. the state should not relinquish control and thereby risk abandoning its responsibilities to wider society. Therefore the role of the state is vital but will always be accompanied by wider societal priorities and related objectives and obligations.

This is why the point of departure between the neo-institutional concept of place-based self-governance through bottom-up polycentric systems and the co-evolutionary hierarchical governance concept (Chapter 4), on which this book is focused, is so significant. The former concept is premised on the assumption (and arguably the ideal) that the state's role should be confined to facilitating and enabling self-governance, whilst the latter concept is premised on the recognition that, in reality, the steering role of the state continues to be of importance, this steer increasingly being achieved through indirect means, i.e. the repositioning and reconfiguring of the role of the state, rather than the retreat of the state, including through the role of legal incentives that both reinforce other incentives and steer them towards the achievement of MPA objectives related to strategic societal priorities. This resonates with wider governance arguments that state steer is still important but that it is achieved by indirect means through coalitions with various non-state actors 'in the shadow of hierarchy' (Bell and Hindmoor, 2009, 71–114).

Implications for empirical research

As was discussed in Chapter 5, the MPAG empirical framework was significantly influenced by the systematic approach developed and applied by neo-institutionalists for analysing governance issues in different case studies. However, the point of departure discussed above means that there are two fundamental differences between case study approaches based on the neo-institutionalist place-based self-governance concept and the MPAG case study approach based on the co-evolutionary hierarchical governance concept.

First, the empirical frameworks and qualitative analysis approaches employed by neo-institutionalist governance analysts are premised on the assumption that the state's role should be confined to facilitating and enabling self-governance. This leads them to focus very specifically, constructively and positively on the role of deliberations amongst local actors. The role of the state, by contrast, is considered positively only in so far as it facilitates and enables self-governance, albeit often on a peripheral basis. More often than not, however, the role of the state is considered negatively, if at all, in terms of undermining bottom-up, self-governance by attempting to impose top-down control. The MPAG empirical framework, on the other hand, is premised on the assumption that the roles of state, market and civil society focused approaches to governance need to be combined, hence the focus on five categories of incentives, all of which are considered positively, including the steering role of the state, markets and people, i.e. the role of the state is positively and constructively embraced by the empirical framework and the underlying theoretical concept, alongside the role of markets and people. This provides for a governance analysis approach that is more comprehensive and that aims to capture the reality that the role of the state needs to be constructively recognised alongside and in combination with the role of other governance approaches.

Second, neo-institutional governance analysts tend to focus on case studies that have been purposefully selected on the basis that they represent self-governance and that they demonstrate the validity of such approaches. This leads towards case studies that are recognised as being successful examples of self-governance. These case studies also tend to be in LECDs and focused on achieving sustainable use objectives related to traditional livelihoods, rather than in a wide range of geopolitical contexts, including MEDCs and focused on achieving both sustainable use and wider conservationist objectives. From an empirical perspective, this is less than ideal as it implies that the sample of case studies will, by design, be biased towards situations that prove the assumption or the hypothesis. By contrast, the MPAs in this study were selected to represent as wide a range of examples and contexts as feasible, i.e. both successful/effective and unsuccessful/ineffective, both LEDCs and MEDCs in a range of geopolitical contexts, both sustainable use and wider conservationist objectives, the latter being inherent in designation as a MPA but the former often also being a priority.

This calls to mind the Amish saying that 'a great deal of what we see depends on what we are looking for'. The MPAG empirical framework and the co-evolutionary hierarchical governance concept on which it is premised has the potential to see how different governance approaches are combined in a wide variety of contexts, rather than only looking for particular things in particular contexts and then arguing that this is what the world should look like. Of course, the empirical use of the MPAG framework is not free of bias and positionality issues: no empirical framework is. It has, however, been designed to try and constructively observe how the full range of governance approaches can be combined in a wide range of MPA case studies. This preliminary study of 20 MPA case studies illustrates the potential of the MPAG framework to provide for qualitative meta-analyses of many more case studies using a more realistic theoretical basis and an empirical framework that can be consistently applied to MPAs in many different contexts. This will enable both the theoretical basis and the empirical framework to be refined, as well as increasing the empirical evidence to inform discussions on governance theories based on realities, rather than on assumptions and ideals.

In the same way that ecological and fisheries scientists have moved on from their earlier studies which were focused on particular species and were confined to limited spatial and temporal scales (Hughes et al., 2005), it is important for governance analysts to move on from their focus on particular categories of incentives in local areas in their focus on place-based self-governance. Faith in the potential of negotiations through horizontal and vertical linkages arguably does not suffice in this respect, therefore constructive recognition of the role of the state in addressing scale challenges and achieving wider-scale societal objectives and obligations becomes of increasing importance.

Overall, the MPAG empirical framework provides for a systematic approach to deconstructing governance issues amongst different case studies in different contexts, providing for consistently structured findings that will enable comparative analyses. Whilst this study has focused on MPAs, the MPAG framework could be adapted to case studies of terrestrial protected areas as well as case studies focused more on sustainable resource use objectives, e.g. fisheries and forestry, the key to this being that we could look for what is happening in reality rather than looking for what we want and expect to find. This research approach has the potential to provide for natural resource governance research, including on MPAs, to be more systematic as well as being more widely focused and unfiltered in its view.

Implications for practice

From the perspective of actors actually engaged in the realities of trying to designate and implement MPAs, the MPAG framework and the findings of its application to the 20 case study analyses can provide considerable guidance. At a general level, it provides for assessments of governance issues in a given MPA that are not constrained by assumptions that self-governance through the participation of local users is the key to effective governance and that the state should only play passive facilitating and enabling roles. Trying to analyse MPA governance from this perspective is often, from a practitioner's view, unrealistic and unfruitful, as governance approaches in a given MPA are rarely confined to user participation and the state rarely confines its roles to passive facilitation and enabling. As the case study discussions in this book illustrate, actual MPA governance invariably combines different governance approaches, including a degree of state control and of user participation, so a governance analysis framework that matches these realities will be more applicable and less constraining.

At a more specific level, the MPAG analysis framework provides a structured approach to analysing governance issues that can be consistently applied to MPAs in a wide range of contexts and that follow a range of MPA governance approaches (government-led, decentralised, etc.). At an even more specific level, the 'menu' of 36 incentives provides a guide for comparing the incentives already applied in a given MPA and for suggesting additional incentives that could be applied to reinforce existing incentives and thereby make the governance approach more effective, including case study examples of how these incentives have been applied and how they reinforce other incentives. The application of the MPAG framework to an increasing number of MPA case studies will provide both for the framework to be refined, particularly in relation to the incentives, and for more examples of different combinations of incentives in different contexts. This will more widely provide for 'good practice' in combinations of different incentives that successfully promote effectiveness to be transferred to other MPAs. The key to the MPAG analysis approach is to demystify and systematise governance analysis for MPA practitioners and to make governance analysis approaches more relevant to the realities of actual MPAs, particularly recognising that providing for the participation of local users *and* for legal incentives to achieve wider conservationist objectives and obligations are not mutually exclusive, i.e. that all five categories of incentives are important to all MPAs, regardless of the governance approach category.

Closing synopsis

There are growing societal concerns about the health of our seas and increasing interest in the potential of MPAs as a key contribution to addressing these concerns. Against this background, this book focuses on the argument that social-ecological resilience can be promoted by institutional diversity. The arguments, framework and findings on which this book is based should help inform discussions on how we can govern MPAs in a way that makes them both effective in achieving their conservationist objectives, particularly in the face of forces that are increasingly driving activities that can challenge the achievement of such objectives, and equitable in addressing related social justice issues. Whilst some divergences between scientists and differences between terrestrial and marine environments pose significant challenges, the need to address the overall challenge of effectively governing MPAs remains paramount, recognising that this will invariably involve addressing basic conflicts underpinned by different value systems.

Given these conflicts, it is argued that the role of the state and of legal incentives, in combination with economic, interpretative, knowledge and participative incentives, needs to be constructively embraced by MPA governance analyses and initiatives. The arguments and findings covered in this book, particularly the MPAG empirical framework and the related co-evolutionary hierarchical governance concept, should enable us to move forward on a more systematic and informed basis in order to steer MPA governance in a manner that meets these challenges, without the constraints of assumptions and ideals related to the concept of self-governance that currently pervade governance analyses. The main aim of this book was stated at the outset to be to explore what the statement that the 'design and management of MPAs must be both top-down and bottom-up' (Kelleher, 1999, xiii) actually means in practice. On the basis of these analyses of 20 case studies it is argued that it means combining a diversity of appropriate incentives from different incentive categories, as diversity is the key to resilience, both of species in ecosystems and incentives in governance systems.

Notes

1 The total MPA coverage figures reported in this book are based on a Policy Brief prepared for the 11th Conference of the Parties to the Convention on Biological Diversity in October 2012 (TNC/UNEP-WCMC, 2012) whilst the total number of MPAs was kindly provided to the author by Dr Mark Spalding, Global Marine Team, The Nature Conservancy. Under the provisions of the United Nations Convention on the Law of the Sea (UNCLOS, 1982), high seas are those beyond successfully claimed exclusive economic zones (EEZs), recognising that not all nations have claimed EEZs out to the limit of 200nm (or the agreed median line where the marine width between nations is less than 400nm), measured from national baselines. Nations can also claim continental shelf areas beyond the 200nm EEZ limit over which they have partial jurisdiction: for non-living and living resource extraction, other than for non-sedentary fisheries. Strictly speaking, areas beyond national jurisdiction, i.e. outside EEZs and claimed continental shelves, that are recognised as common heritage are known as 'the Area', but they are also commonly referred to as 'high seas', even though actual high seas include successfully claimed continental shelf extensions that are under partial national jurisdiction. This book will employ the commonly used term 'high seas' to refer to areas beyond national jurisdiction, recognising that legally such areas are known as 'the Area'. It is also worth noting that the World Database on Protected Areas (IUCN/UNEP-WCMC, 2011) uses the median line or the geographical 200nm limit to define areas under national jurisdiction (39 per cent total marine area), regardless of the fact that some nations have not fully claimed EEZs out to the median line or 200nm limit, and that some nations have claimed continental shelves beyond 200nm, these being under partial national jurisdiction. This should be recognised when considering the MPA coverage figures.

2 Strictly speaking, this specific definition of MPAs has been superseded by the new definition that applies to all protected areas: 'A clearly defined geographical space, recognised, dedicated and managed, through legal or other effective means, to achieve the long-term conservation of nature with associated ecosystem services and cultural values' (Dudley, 2008, 8). This was intended to demarcate between protected areas focused on achieving conservation objectives and those sites 'where the primary purpose is extractive uses, i.e. fisheries management areas' (Dudley, 2008, 56). However, the new IUCN protected area classification scheme added the new category of 'protected area with sustainable use of natural resources' (VI) which could feasibly include fisheries management areas, provided the primary objective is 'to protect natural ecosystems and use natural resources sustainably' and fisheries management includes measures to reduce fishing impacts on habitats, e.g. bans on destructive gears, and species, e.g. bycatch control. The original specific MPA definition (1991) is still widely used, including by the IUCN. It is usually attributed to the more recent MPA guidance (Kelleher, 1999), but the original source of this definition is attributed here.

3   Encyclopædia Britannica Online Academic Edition. Encyclopædia Britannica Inc., 2012. http://www.britannica.com/EBchecked/topic/1669736/trophic-cascade, accessed 06 July 2012.

4   The primary MPA goal agreed by the IUCN: 'To provide for the protection, restoration, wise use, understanding and enjoyment of the marine heritage of the world in perpetuity through the creation of a global representative system of MPAs and through the management in accordance with the principles of the World Conservation Strategy of human activities that use or affect the marine environment' (Resolution 17.38, 17th General Assembly of the IUCN, 1988, reported by Kelleher and Kenchington, 1991, xx).

5   The no-take MPA coverage was updated from Wood et al. (2008) by adding the no-take area of eight very large MPAs designated since the end of 2006: Chagos Archipelago MPA, British Indian Ocean Territory (640,000 km^2 no-take), South East Commonwealth Marine Reserve Network (154,435 km^2 no-take); Phoenix Islands Protected Areas, Kiribati (15,800 km^2 no-take); Prince Edward Islands MPA, South Africa (61,415 km^2 no-take); South Orkneys MPA, British Antarctic Territory (93,787 km^2 no-take); Motu Motiro Hiva Marine Park, Chile (150,000 km^2 no-take); Coral Sea Commonwealth Marine Reserve, Australia (502,654 km^2 no-take) and South Georgia and South Sandwich Islands MPA (20,000 km^2 no-take). The total no-take MPA coverage (~1.93 million km^2) represents 1.37 per cent of the marine area under national jurisdiction and 0.54 per cent of the global marine area, up from 0.2 per cent and 0.08 per cent respectively since 2006 (Wood et al., 2008). It is important to recognise that the vast majority (85 per cent) of this no-take MPA coverage is attributable to only these eight very large no-take MPA designations in remote areas. It is also important to recognise that whilst these MPAs are officially designated as no-take, the reality is that enforcing such bans on all local and incoming fishing vessels over such vast, remote areas raises major surveillance and enforcement challenges, particularly given the growing scale of illegal, unreported and unregulated (IUU) fishing, and it is highly debatable whether these areas are actually 'no-take'.

6   Given the combination of high scientific uncertainty, high societal stakes and a high diversity or plurality of value perspectives, research related to MPAs could be considered as representing post-normal science (Ravetz, 1999), in the context of which it is to be expected that scientists will depart from a purely objective, positivist approach as wider societal values are integral to post-normal science, as previously discussed by Jones (2001). The differences between positivist and the post-normal views on the role of science in research related to MPAs can also be considered in terms of the distinction between Mode 1 and Mode 2 science, the former characterised as being reductive, intra-disciplinary and applied, but not societally accountable/inclusive, and the latter as holistic, trans-disciplinary and carried out in collaboration with society in the context of application, with which it co-evolves (Gibbons et al., 1994). Mode 2 science recognises that uncertainties will proliferate rather than being progressively eradicated, therefore they should be accommodated on a precautionary basis rather than feared, including new societal (not just scientific) innovations to cope with uncertainties (Nowotony et al., 2001), as previously discussed by Jones (2007). The divergences between the quest for a positivist, Mode 1 approach by many fisheries scientists and the acceptance of a post-normal, Mode 2 approach, particularly where the former approach is not feasible and/ or appropriate, by many marine ecologists are a key factor behind many of the divergent views between advocates of CFMAs and of MPAs discussed in Chapter 3. In this sense, the governance of MPAs could also be considered to represent a series of 'wicked problems' (Jentoft and Chuenpagdee, 2009).

7   The term 'Malthusian analysis of despair' is related to Thomas Malthus' *Essay on the Principle of Population*, first published in 1798. This essay argued that the world's human

population would grow at a much faster rate than agricultural productivity, resulting in mass starvation and much human misery.

8 This quote is from a MPA News article (vol 14, no 2, p. 8, July/August 2012) highlighting, inter alia, the launch of new guidelines on applying IUCN protected area categories to MPAs (IUCN, 2012). It goes on to clarify that 'if marine areas involve extractive uses and have no defined long-term goals of conservation and ocean recovery, they should not be considered as MPAs'. It also quotes Dan Laffoley (Marine Vice-Chair, IUCN World Commission on Protected Areas): 'It is time to stop pretending more of the ocean is protected than actually is'. This is arguably a reaction to the increasing interest in Category V and VI MPAs that provide for sustainable fishing activities, as discussed in Chapter 2, and to the provision for fishing activities in other categories of MPAs which should be no-take, particularly Ia, Ib and II. It is certainly an indication that the divergence between fisheries scientists and marine ecologists on the need for and role of no-take MPAs remains a significant one.

9 Proprietorship is a key concept in community-based conservation, which means 'sanctioned use rights, including the right to determine the mode and extent of management and use, rights of access and inclusion, and the right to benefit fully from use and management ... Delegation or proprietorship over natural resources to communities requires the state to relinquish considerable authority and responsibility'. (Murphree, 1994, 405–406). Appropriation means, in this context, that the state has removed community proprietorship and assumed the role of proprietor through the increased centralisation of control.

10 It is interesting but worrying to note that there is a perverse symmetry with respect to equity, in that disadvantaged sectors of local communities may become marginalised from decision-making, as well as bearing an unfair proportion of the costs and being denied access to benefits, through both top-down (risks of imposition associated with unholy alliances between corporate tourism interests, the state and conservation NGOs) and bottom-up approaches (risks of parochialism associated with elite capture).

11 Such financial incentives are provided for by the CBD, Article 11 of which includes obligations to adopt economically and socially sound measures that act as incentives for the conservation and sustainable use of biodiversity, the GEF being a facility for countries, particularly MEDCs, to contribute to such incentives under the CBD.

12 The CBD also recognises that biodiversity represents a 'common concern of humankind' (Preamble) but the provision for sovereign rights under Article 3 prevails. The wider Common Heritage of Mankind principle is recognised in the preamble to the Convention Concerning The Protection of the World Cultural and Natural Heritage (1972): 'Considering that parts of the cultural or natural heritage are of outstanding interest and therefore need to be preserved as part of the world heritage of mankind as a whole.' Such interest includes sites which 'contain the most important and significant natural habitats for in-situ conservation of biological diversity, including those containing threatened species of outstanding universal value from the point of view of science or conservation' (Selection Criterion 10, Operational Guidelines, 2005, see http://whc.unesco.org/en/criteria).

13 In particular, the application of co-management and adaptive co-management approaches to a MPA in Honduras is critically analysed in great detail by Bown et al. (2012), with a focus on the lack of community participation and empowerment provided for by the state agency. The conclusions, however, are limited to stating that more participation and empowerment is needed, along with other ways in which adaptive co-management 'might be improved to approximate more closely to the ideal of ACM', which essentially complies with the neo-institutional ideals discussed in Chapter 4.

14 The term 'state capacity' is used in this study to: (a) reflect the fact that the governance capacity metrics evaluate the quality of governance mainly from a government or state

function perspective; and (b) avoid confusion with our use of the term 'governance' to include a wider range of governance approaches other than government or state functions.

15 Web links for further information and national data on these metrics: CIA World Factbook – https://www.cia.gov/library/publications/the-world-factbook; state capacity – www.govindicators.org; human development index – http://hdr.undp.org/en/statistics/hdi.

16 Whilst there are differences between policy and law, they are generally closely related, if not integrated. Laws are usually passed through a national democratic government process to achieve specific societal objectives, sometimes also to fulfil legal obligations imposed on the national government, and subject to enforcement through surveillance (involving police, wardens, officers, etc.), political and judicial systems. Policies are the official decisions and guidance that support the implementation of one or more laws and the fulfilment of the same societal objectives. Non-compliance with policies can be addressed by official support, including funding, and political/bureaucratic pressure to comply and, where non-compliance with a policy equates to non-compliance with a law, ultimately by recourse to judicial enforcement. Most policies are related to laws, policies that are not related to laws tending to be less binding and influential in that they tend to reflect a lower government priority, so there may be less official support and political/bureaucratic pressure to comply, and there is not the legal incentive to comply in order to avoid prosecution. The national capacity to produce and enforce laws and policies is indicated by the state capacity metric described in Chapter 5.

17 The recognition of the instrumental role of markets brought the following cautionary quote to mind: 'For all their power and vitality, markets are only tools. They make a good servant but a bad master and a worse religion' (Hawken et al., 1999, 261).

18 Similar concerns that compensation can escalate are also emerging in relation to the Phoenix Islands Protected Area (PIPA) 'reverse fishing licence' scheme, as an increase in the value of tuna, and therefore the potential income from selling fishing licences to foreign vessels, is leading to extra compensation, in the form of donations to a trust fund from BINGOs, being requested by the Kiribati government to close further areas to tuna fishing, coupled with concerns that the same number of licences will be sold but that effort will simply be displaced to non-designated areas of the Phoenix Islands EEZ (Anon., 2013).

19 This recognition of the particular importance of legal incentives is why participative incentives related to social capital were not included as the foundations of the building in this metaphor, as even the firmest of foundations will not prevent the collapse of a governance framework that lacks the structural stability of legal reinforcements.

References

Abesamis, R.A., Alcala, A.C. and Russ, G.R. (2006) 'How much does the fishery at Apo Island benefit from spillover of adult fish from the adjacent marine reserve?', *Fisheries Bulletin*, 104, 360–375.

Adams, W.M. (2004) *Against Extinction: the story of conservation*, Earthscan, London.

Adams, W.M. and Hutton, J. (2007) 'People, parks and poverty: political ecology and biodiversity conservation', *Conservation and Society*, 5, 147–183.

Adams, W.M., Aveling, R., Brockington, D., Dickson, B., Elliott, J., Hutton, J., Roe, D., Vira, B. and Wollmer, W. (2004) 'Biodiversity conservation and the eradication of poverty', *Science*, 306, 1146–1149.

Agardy, T. (1994) *The Science of Conservation in the Coastal Zone: new insights on how to design, implement and monitor marine protected areas*, IUCN, Gland, Switzerland.

Agrawal, A. (2001) 'Common property institutions and sustainable governance of resources', *World Development*, 29, 1649–1672.

Airamé, S., Dugan, J.E., Lafferty, K.D., Leslie, H., McArdle, D.A. and Warner, R.R. (2003) 'Applying ecological criteria to marine reserve design: a case study from the California Channel Islands', *Ecological Applications*, 13, 1 (Supplement), S170–S184.

Alcala, A.C., Russ, G.R., Maypa, A.P. and Calumpong, H.P. (2005) 'A long-term, spatially replicated experimental test of the effect of marine reserves on local fish yields', *Canadian Journal of Fisheries and Aquatic Sciences*, 62, 98–108.

Alder, J. (1996) 'Have tropical marine protected areas worked? An analysis of their success', *Coastal Management*, 24, 97–114.

Anon. (2002) 'Results from the Reader Challenge: which MPA is the oldest?', *MPA News*, December 2001/January 2002, 6, 4, 5–6.

——(2008) 'Link by Link – a short history of modern finance', *The Economist*, 389, 96–97.

——(2013) 'The reverse fishing license mechanism for Kiribati's Phoenix Islands Protected Area: an experiment in MPA financing', *MPA News*, July/August 2013, 15, 1, 1–3.

Armitage, D. and Plummer, R. (2010) 'Adapting and transforming: governance for navigating change', pp. 287–302 in D. Armitage and R. Plummer (eds) *Adaptive capacity and environmental governance*, Springer, New York.

Arnason, R., Kelleher, K. and Willman, F. (2008) *The Sunken Billions: the economic justification for fisheries reform*, The World Bank, Washington DC/FAO, Rome.

Arrieta, J.M., Arnuad-Haound, S. and Duarte, C.M. (2010) 'What lies underneath: conserving the oceans' genetic resources', *Proceedings of the National Academy of Sciences*, 107, 43, 18318–18324.

Asafu-Adjaye, J. (2000) 'Customary marine tenure systems and sustainable fisheries management in Papua New Guinea', *International Journal of Social Economics*, 27, 917-927.

Babcock, R.C., Shears, N.T., Alcala, A.C., Barrett, N.S., Edgar, G.J., Lafferty, K.D., McClanahan, T.R. and Russ G.R. (2010) 'Decadal trends in marine reserves reveal differential rates of change in direct and indirect effects', *Proceedings of the National Academy of Sciences*, 107, 43, 18256-18261.

Ballantine, W.J. (1991) *Marine Reserves for New Zealand*, Leigh Laboratory Bulletin No. 25, University of Auckland, Auckland.

—(1997) 'Design principles for systems of "no-take" marine reserves', in *Workshop on the Design and Monitoring of Marine Reserves, February 18–20*, Fisheries Centre, University of British Columbia, Canada.

—(2002) 'MPA Perspective: MPAs improve general management, while marine reserves ensure conservation', *MPA News* 4, 1, 5.

Barry, D. and Oelschlaeger, M. (1996) 'A science for survival: values and conservation biology', *Conservation Biology*, 10, 905-911.

Bascompte, J., Melián, C.J. and Sala, E. (2005) 'Interaction strength combinations and the overfishing of a marine food web', *Proceedings of the National Academy of Sciences*, 102, 5443-5447.

Bates, A.E., Barrett, N.S., Stuart-Smith, R.D., Holbrook, N.J., Thompson, P.A. and Edgar, G.J. (2014) 'Resilience and signatures of tropicalization in protected reef fish communities', *Nature Climate Change*, 4, 62-67

Baum, J.K., Myers, R.A., Kehler, D.G., Worm, B., Harley, S.J. and Doherty, P.A. (2003) 'Collapse and conservation of shark populations in the Northwest Atlantic', *Science*, 299, 389-392.

Beaumont, N.J., Austen, M.C., Atkins, J.P., Burdon, D., Degraer, S., Dentinho, T.P., Derous, S., Holm, P., Horton, T., van Ierland, E., Marboe, A.H., Starkey, D.J., Townsend, M. and Zarzycki, T. (2007) 'Identification, definition and quantification of goods and services provided by marine biodiversity: implications for the ecosystem approach', *Marine Pollution Bulletin*, 54, 253-265.

Beddington, J.R., Agnew, D.J. and Clark, C.W. (2007) 'Current problems in the management of fisheries', *Science*, 316, 1713-1716.

Bell, S. and Hindmoor, A. (2009) *Rethinking Governance: the centrality of the state in modern society*, Cambridge University Press, Cambridge, UK.

Bellwood, D.R., Hughes, T.P., Folke, C. and Nyström, M. (2004) 'Confronting the coral reef crisis', *Nature*, 429, 827-833.

Berkeley, S.A., Chapman, C. and Sogard, S.M. (2004) 'Maternal age as a determinant of larval growth and survival in a marine fish, Sebastes melanops', *Ecology*, 85, 1258-1264.

Berkes, F. (2002) 'Cross-scale institutional linkages: perspectives from the bottom up', pp. 293-321 in E. Ostrom, T. Dietz, N. Dolšak, P.C. Stern, S. Sonich and E.U. Weber (eds) *The Drama of the Commons*, National Academy Press, Washington DC.

—(2007) 'Community-based conservation in a globalized world', *Proceedings of the National Academy of Sciences*, 104, 15188-15193.

—(2011) 'Restoring unity: the concept of marine social-ecological systems', pp. 9-28 in R. Ommer, I. Perry, K.L. Cochrane and P. Curry (eds) *World Fisheries: A Social-Ecological Analysis*, Wiley, New Jersey.

Berkes, F. and Folke, C. (eds) (1998) *Linking Social and Ecological Systems*, Cambridge University Press, Cambridge, UK.

Berkes, F. and Seixas, C.S. (2008) 'Community-based enterprises: the significance of horizontal and vertical institutional linkages'. Paper presented at the 12th Biennial Conference of the International Association for the Study of the Commons - *Governing Shared Resources: connecting local experience to global challenges*. July 2008, University of Gloucestershire, UK.

Blaustein, R.J. (2007) 'Protected areas and equity concerns', *Bioscience*, 57, 216-221.

Blount, B.G. and Pitchon, A. (2007) 'An anthropological research protocol for marine protected areas: creating a niche in a multidisciplinary hierarchy', *Human Organization*, 66, 103-111.

Bodin, O., Crona, B. and Ernstsen, H. (2006) 'Social networks in natural resource management: what is there to learn from a structural perspective', *Ecology and Society*, 11, 2, response 2.

Borrini-Feyerabend, G. (1999) 'Collaborative management of protected areas', pp. 225-234 in S. Stolton and N. Dudley (eds) *Partnerships for Protection: new strategies for planning and management for protected areas*, Earthscan, London.

Borrini-Feyerabend, G., Pimbert, M., Farvar, M.T., Kothari, A. and Renard, Y. (2007) *Sharing Power: a global guide to collaborative management of natural resources*, Earthscan, London.

Botsford, L.W., Castilla, J.C. and Peterson, C.H. (1997) 'The management of fisheries and marine ecosystems', *Science*, 277, 509-515.

Bottema, M.J.M. and Bush, S.R. (2012) 'The durability of private sector-led marine conservation: a case study of two entrepreneurial marine protected areas in Indonesia', *Ocean & Coastal Management*, 61, 38-48.

Bown, N., Gray, T. and Stead, S.M. (2013) *Contested Forms of Governance in Marine Protected Areas: a study of co-management and adaptive co-management*, Routledge, Abingdon, Oxfordshire.

Brailovskaya, T. (1998) 'Obstacles to protecting marine biodiversity through marine wilderness preservation: examples from the New England region', *Conservation Biology*, 12, 1236-1240.

Branch, T.A., Jensen, O.P., Ricard, D., Ye, Y. and Hilborne, R. (2011) 'Contrasting global trends in marine fishery status obtained from catches and from stock assessments', *Conservation Biology*, 25, 4, 777-786.

Brockington, D. (2002) *Fortress Conservation: the preservation of the Mkomazi Game Reserve, Tanzania*, James Currey, Oxford.

Brockington, D., Duffy, R. and Igoe, J. (2008) *Nature Unbound: conservation, capitalism and the future of protected areas*, Earthscan, London.

Brondizio, E.S., Ostrom, E. and Young, O.A. (2009) 'Connectivity and the governance of multilevel social-ecological systems: the role of social capital', *Annual Review of Environment and Resources*, 34, 252-278.

Brotz, L., Cheung, W.W.L., Kleisner, K., Pakhomov, E. and Pauly, D. (2012) 'Increasing jellyfish populations: trends in large marine ecosystems', *Hydrobiologia*, 690, 3-20.

Büscher, B. and Whande, W. (2007) 'Whims of the winds of time? Emerging trends in biodiversity conservation and protected area management', *Conservation and Society*, 5, 22-43.

Callicot, J.B. (1991) 'Conservation ethics and fishery management', *Fisheries*, 16, 2, 22-28.

Campbell, S.J., Kartawijaya, T., Yulianto, I., Prasetia, R. and Clifton, J. (2013) 'Co-management approaches and incentives to improve management effectiveness in the Karimunjawa National Park, Indonesia', *Marine Policy*, 41, 72-79.

Cannon, T. and Müller-Mahn, D. (2010) 'Vulnerability, resilience and development discourses in context of climate change', *Natural Hazards*, 55, 621-635.

Carr, M. and Raimondi, P.T. (1999) 'Marine protected areas as a precautionary approach to management', *California Cooperative Oceanic Fisheries Investigations Reports*, 40, 71-76.

Cash, D.W., Adger, W.N., Berkes, F., Garden, P., Lebel, L., Olsson, P., Pritchard, L. and Young, O. (2006) 'Scale and cross-scale dynamics: governance and information in a multi-level world', *Ecology and Society*, 11, 2, article 8.

Casini, M., Lövgren, J., Hjelm, J., Cardinale, M., Molinero, J-C. and Kornilovs, G. (2008) 'Multi-level trophic cascades in heavily exploited open marine ecosystems', *Proceedings of the Royal Society B*, 275, 1793-1801.

Causey, B.D. (1995) Enforcement in marine protected areas, pp. 119-148 in S. Gubbay (ed) *Marine Protected Areas: principles and techniques for management*, Chapman & Hall, London.

Caveen, A.J., Sweeting, C.J., Willis, T.J. and Polunin, N.V.C. (2012) 'Are the scientific foundations of temperate marine reserves too warm and hard?', *Environmental Conservation*, 39, 199-203.

Chambers, R. (1983) *Rural Development: putting the last first*, Longman, Harlow.

Chapin, M. (2004) 'A challenge to conservationists', *World Watch Magazine*, November/December 2004, 17-31.

Charette, M.A. and Smith, W.H.F. (2010) 'The volume of Earth's ocean', *Oceanography*, 23, 104-106.

Christie, P. and White, A.T. (1997) 'Trends in development of coastal area management in tropical countries: from central to community orientation', *Coastal Management*, 25, 155-181.

Chuenpagdee, R., Pascual-Fernández, J.J., Szeliánszky, M., Alegret, J.L., Fraga, J. and Jentoft, S. (2013) 'Marine Protected Areas: rethinking their inception', *Marine Policy*, 39, 234-240.

Cinner, J.A. and Aswani, S. (2007) 'Integrating customary management into marine conservation', *Biological Conservation*, 140, 201-216.

Cleaver, F. (2001) 'Institutions, agency and the limitations of participatory approaches to development', pp. 36-55 in B. Cooke and U. Kothari (eds) *Participation: the new tyranny?*, Zed Books, London/New York.

Clifton, J. (2013) 'Refocusing conservation through a cultural lens: improving governance in the Wakatobi National Park, Indonesia', *Marine Policy*, 41, 80-86.

Colwell, S. (1997) 'Entrepreneurial marine protected areas: small-scale, commercially supported coral reef protected areas', pp. 110-114 in M.E. Haziolos, A.J. Hooten and M. Fodor (eds) *Coral Reefs: challenges and opportunities for sustainable management*, World Bank, Washington DC.

Condon, R.H., Duarte, C.M., Pitt, K.A., Kelly, L., Robinson, K.L., Lucas, C.H., Sutherland, K.R., Mianzan, H.W., Bogeberg, M., Purcell, J.E., Decker, M.B., Uye, S., Madin, L.P., Brodeur, R.D., Haddock, S.H.D., Malej, A., Parry, G.D., Eriksen, E., Quiñones, J., Acha, M., Harvey, M., Arthur, J.M. and Graham, W.M. (2013) 'Recurrent jellyfish blooms are a consequence of global oscillations', *Proceedings of the National Academy of Sciences*, 110, 1000-1005.

Conover, D.O. and Munch, S.B. (2002) 'Sustaining fisheries yields over evolutionary time scales', *Science*, 297, 94-96.

Cooke, A.J., Polunin, N.V.C. and Moce, K. (2000) 'Comparative assessment of stakeholder management in traditional Fijian fishing-grounds', *Environmental Conservation*, 27, 291-299.

Cooke, B. and Kothari, U. (2001) 'The case for participation as tyranny', pp. 1-15 in B. Cooke and U. Kothari (eds) *Participation: the new tyranny?*, Zed Books, London/New York.

Cooke, S.J., Danylchuck, A.J., Danylchuck, S.E., Suski, C.D. and Goldberg, T.L. (2006) 'Is catch-and-release recreational angling compatible with no-take marine protected areas?', *Ocean & Coastal Management*, 49, 342-354.

Corbera, E. (2012) 'Problematizing REDD+ as an experiment in payments for ecosystem services', *Current Opinion in Environmental Sustainability*, 4, 612-619.

Corbera, E. and Schroeder, H. (2011) 'Governing and implementing REDD+', *Environmental Science and Policy*, 14, 89-99.

Costanza, R., d'Arge, R., de Groot, R.S., Farber, S., Grasso, M., Hannon, B., Limburg, K., Naeem, S., O'Neill, R., Paruelo, J., Raskin, R., Sutton, P. and van den Belt, M. (1997) 'The value of the world's ecosystem services and natural capital', *Nature* 387, 253-260.

Costello, C., Gaines, S.D. and Lynham, J. (2008) 'Can catch shares prevent fisheries collapse?', *Science*, 321, 1678-1681.

Costello, C., Ovando, D., Hilborn, R., Gaines, S.D., Deschenes, O. and Lester, S.E. (2012) 'Status and solutions for the world's unassessed fisheries', *Science*, 338, 517-520.

Crabbe, M.J.C., Martinez, E., Garcia, C., Chub, J., Castro, L. and Guy, J. (2010) 'Is capacity building important in policy development for sustainability? A case study using action plans for sustainable marine protected areas in Belize', *Society and Natural Resources*, 23, 181-190.

Crec'hriou, R., Alemany, F., Roussel, E., Chassanite, A., Marinaro, J.Y., Mader, J., Rochell, E. and Planes, S. (2010) 'Fisheries replenishment of early life taxa: potential export of fish eggs and larvae from a temperate marine protected area', *Fisheries Oceanography*, 19, 135-150.

Crowder, L.B., Lyman, S.J., Figueira, W.F. and Priddy, J. (2000) 'Source-sink population dynamics and the problem of siting marine reserves', *Bulletin of Marine Science*, 66, 799-820.

Crowder, L.B. and Norse, E.A. (2008) 'Essential ecological insights for marine ecosystem-based management', *Marine Policy*, 32, 772-778.

Croxall, J.P., Butchart, S.H.M., Lascelles, B,. Strattersfield, A.J., Sullivan, B., Symes, A. and Taylor, P. (2012) 'Seabird conservation status, threats and priority actions: a global assessment', *Bird Conservation International*, 22, 1-34.

Dalton, R. (2004) 'Net losses pose extinction risk for porpoise', *Nature*, 429, 590.

——(2005) 'Is this any way to save a species?', *Nature*, 436, 14-16.

——(2010) 'Reserves 'win-win' for fish and fishermen', *Nature*, 463, 1007.

Daskalov, G.M., Grishin, A.N., Rodionov, S. and Mihneva, V. (2007) 'Trophic cascades triggered by overfishing reveal possible mechanisms of ecosystem regime shifts', *Proceedings of the National Academy of Sciences*, 104, 10518-10523.

Davidson, O.G. (1998) *The Enchanted Braid: coming to terms with nature on the coral reef*, John Wiley, New York.

Davis, D. and Tisdell, C. (1995) 'Recreational scuba-diving and carrying capacity on marine protected areas', *Ocean & Coastal Management*, 26, 19-40.

Day, J., Dudley, N., Hockings, M., Holmes, G., Laffoley, D., Stolton, S. and Wells, S. (2012) *Guidelines for applying the IUCN Protected Area Management Categories to Marine Protected Areas*, IUCN, Gland, Switzerland.

Day, J.C. and Dobbs, K. (2013) 'Effective governance of a large and complex cross-jurisdictional MPA: Australia's Great Barrier Reef', *Marine Policy*, 41, 14-24.

Dayton, P.K., Tegner, M.J., Edwards, P.B. and Riser, K.L. (1998) 'Sliding baselines, ghosts, and reduced expectations in kelp forest communities', *Ecological Applications*, 8, 2, 309-322.

Dayton, P.K., Thrush, S.F., Agardy, M.T. and Hofman, R.J. (1995) 'Environmental effects of marine fishing', *Aquatic Conservation – Marine and Freshwater Ecosystems*, 5, 3, 205-232.

Dearden, P., Bennett, M. and Johnston, J. (2005) 'Trends in global protected area governance, 1992-2002', *Environmental Management*, 36, 89-100.

Depondt, F. and Green, E. (2006) 'Diving user fees and the financial sustainability of marine protected areas: opportunities and impediments', *Ocean & Coastal Management*, 49, 188-202.

De Groot, J. and Bush, S.R. (2010) 'The potential for dive tourism led entrepreneurial marine protected areas in Curacao', *Marine Policy*, 34, 1051-1059.

De Santo, E.M. (2012) 'From paper parks to private conservation: the role of NGOs in adapting marine protected area strategies to climate change', *Journal of International Wildlife Law & Policy*, 15, 25-40.

——(2013a) 'Missing marine protected area (MPA) targets: how the push for quantity over quality undermines sustainability and social justice', *Journal of Environmental Management*, 124, 137-146.

——(2013b) 'The Darwin Mounds Special Area of Conservation, United Kingdom', *Marine Policy*, 41, 25-32.

De Santo, E.M., Jones, P.J.S and Miller, A. (2011) 'Fortress conservation at sea: a commentary on the Chagos MPA', *Marine Policy*, 35, 258-260.

Dietz, T., Dolšak, N., Ostrom, E. and Stern, P.C. (2002) 'The Drama of the Commons', pp. 3-35 in E. Ostrom, T. Dietz, N. Dolšak, P.C. Stern, S. Sonich and E.U. Weber (eds) *The Drama of the Commons*, National Academy Press, Washington, DC.

Dight, I.J. (1995) 'Understanding larval dispersal and habitat connectivity in tropical marine systems: a tool for management', pp. 41-45 in T. Agardy (ed.) *The Science of Conservation in the Coastal Zone: new insights on how to design, implement and monitor marine protected areas*, IUCN, Gland, Switzerland.

Dikou, A. (2010) 'Ecotourism in marine protected areas: development, impacts and management – a critical review', pp. 1-42 in A. Kraus and E. Weir (eds) *Ecotourism: management, development and impact*, Nova Science Publishers, New York.

Done, T.J. (1992) 'Phase shifts in coral reef communities and their ecological significance', *Hydrobiologia*, 247, 121-132.

Douvere, F. (2008) 'The importance of marine spatial planning in advancing ecosystem-based sea use management', *Marine Policy*, 32, 762-771.

Douvere, F. and Ehler, C.N. (2010) 'The importance of monitoring and evaluation in adaptive maritime spatial planning', *Journal of Coastal Conservation*, 15, 305-311.

Dreyfus, H.L. and Rabinow, P. (1982) cited in Pinkus, J. (1996) *Foucault*, http://www.massey.ac.nz/~alock/theory/foucault.htm, accessed 28 January 2013.

Dryzek, J.S. (1987) *Rational Ecology: environment and political economy*, Basil Blackwell, New York.

——(2013, Third Edition) *The Politics of the Earth: environmental discourses*, Oxford University Press, Oxford/New York.

Dudley, N. (2008) *Guidelines for Applying Protected Area Management Categories*, IUCN, Gland, Switzerland.

Duit, A. (2012) 'Adaptive capacity and the ecostate', pp. 127–147 in E. Boyd, and C. Folke, (eds) *Adapting Institutions: governance, complexity and social-ecological resilience*, Cambridge University Press, Cambridge, UK.

Dulvy, N.K., Freckleton, R.P. and Polunin, N.V.C. (2004) 'Coral reef cascades and the indirect effects of predator removal by exploitation', *Ecology Letters*, 7, 410–416.

Dulvy, N.K. Sadovy, Y. and Reynolds, J.D. (2003) 'Extinction vulnerability in marine populations', *Fish and Fisheries*, 4, 25–64.

Dygico, M., Songco, A., White, A.T. and Green, S.J. (2013) 'Achieving MPA effectiveness through application of responsive governance incentives in the Tubbataha reefs', *Marine Policy*, 41, 87–94.

Ebbesson, J. (2010) 'The rule of law in governance of complex socio-ecological changes', *Global Environmental Change*, 20, 414–422.

Ebrahim, A. (2003) 'Making sense of accountability: conceptual perspectives for northern and southern nonprofits', *Nonprofit Management and Leadership*, 14, 191–212.

Edwards, V.M. and Steins, N.A. (1999) 'Special issue introduction: the importance of context in common pool resource research', *Journal of Environmental Policy & Planning*, 1, 195–204.

Ehler, C. and Douvere, F. (2007) *Visions for a sea change*, Report of the First International Workshop on Marine Spatial Planning, Intergovernmental Oceanographic Commission and Man and the Biosphere Programme, IOC Manual and Guides, 48, UNESCO, Paris.

Ellis, E.A. and Porter-Bolland, L. (2008) 'Is community-based forest management more effective than protected areas? A comparison of land use/land cover change in two neighbouring study areas of the Central Yucatan Peninsula, Mexico', *Forest Ecology and Management*, 256, 1971–1983.

Elmqvist, T., Folke, C., Nyström, M., Peterson, G., Bengtsson, J., Walker, B. and Norberg, J. (2003) 'Response diversity, ecosystem change, and resilience', *Frontiers in Ecology and the Environment*, 1, 488–494.

Essington, T.E. and Punt, A.E. (2011) 'Editorial – Implementing ecosystem-based fisheries management: advances, challenges and emerging tools', *Fish and Fisheries*, 12, 123–124.

Estes, J.A. and Palmissano, J.F. (1974) 'Sea otters: their role in structuring nearshore communities', *Science*, 185, 1058–1060.

Estes, J.A., Terborgh, J., Brashares, J.S., Power, M.E., Berger, J., Bond, W.J., Carpenter, S.R., Essington, T.E., Holt, R.D., Jackson, J.B.C., Marquis, R.J., Oksanen, L., Oksanen, T., Paine, R.T., Pikitch, E.K., Ripple, W.J., Sandin, S.A., Scheffer, M., Schoener, T.W., Shurin, J.B., Sinclair, A.R.E., Soulé, M.E., Virtanen, R. and Wardle, D.A. (2011) 'Trophic downgrading of Planet Earth', *Science*, 333, 301–306.

Estes, J.A., Tinker, M.T., Williams, T.M. and Doak, D.F. (1998) 'Killer whale predation on sea otters linking oceanic and nearshore ecosystems', *Science*, 282, 473–476.

FAO (2010) *The State of the World Fisheries and Aquaculture 2010*, Food and Agriculture Organisation, Rome.

——(2012) *The State of the World Fisheries and Aquaculture 2012*, Food and Agriculture Organisation, Rome.

Fabinyi, M. (2008) 'Dive tourism, fishing and marine protected areas in the Calamianes Islands, Philippines', *Marine Policy*, 32, 898–904.

Fanning, L., Mahon, R., McConney, P., Angulo, J., Burrows, F., Chakalall, B., Gil, D., Haughton, M., Heileman, S., Martíinez, S., Ostine, L., Oviedo, A., Parsons, S., Phillips, T., Arroya, C.S., Simmons, B. and Toro, C. (2007) 'A large marine ecosystem governance framework', *Marine Policy*, 31, 434–443.

Fenner, D. (2012) 'Challenges for managing fisheries on diverse coral reefs', *Diversity*, 4, 105-160.

Fernandes, L., Day, J., Lewis, A., Slegers, S., Kerrigan, B., Breen, D., Cameron, D., Jago, B., Hall, J., Lowe, D., Innes, J., Tanzer, J., Chadwick, V., Thompson, L., Gorman, K., Simmons, M., Barnett, B., Sampson, K., De'ath, G., Mapstone, B., Marsh, H., Possingham, H., Ball, I., Ward, T., Dobbs, K., Aumend, J., Slater, D. and Stapleton, K. (2005) 'Establishing representative no-take areas in the Great Barrier Reef: large-scale implementation of theory on marine protected areas', *Conservation Biology*, 19, 6, 1733-1744.

Folke, C., Hahn, T., Olsson, P. and Norberg, J. (2005) 'Adaptive governance of social-ecological systems', *Annual Review of Environment and Resources*, 30, 441-473.

Frank, K.T., Petrie, B., Choi, J.S. and Leggett, W.C. (2005) 'Trophic cascades in a formerly cod dominated ecosystem', *Science*, 308, 1621-1623.

Fujita, R. and Bonzon, K. (2005) 'Rights-based fisheries management: an environmentalist perspective', *Reviews in Fish Biology and Fisheries*, 15, 309-312.

GEF/UNDP (Global Environment Facility/United Nations Development Programme) (2012) *Catalysing Ocean Finance, volume 1: Transforming Markets to Restore and Protect the Global Ocean*, UNDP, New York.

GOL (Global Ocean Legacy, 2011) *Worlds Marine Scientists Call for Large Scale National Parks at Sea.* http://www.pewenvironment.org/news-room/press-releases/worlds-marine-scientists-call-for-large-scale-national-parks-at-sea-8589935354, accessed 28 January 2012.

Garcia, S.M. and Rosenberg, A.A. (2010) 'Food security and marine capture fisheries: characteristics, trends, drivers and future perspectives', *Philosophical Transactions of the Royal Society B*, 365, 2869-2880.

Garmestani, A.S. and Benson, M.H. (2013) 'A framework for resilience-based governance of social-ecological systems', *Ecology and Society*, 18, 1, article 9.

Gelcich, S., Godoy, N., Prado, L. and Castilla, J.C. (2008) 'Add-on conservation benefits of marine territorial user rights fisheries policies in Central Chile', *Ecological Applications*, 18, 273-281.

Gell, F.R. and Roberts, C.M. (2003) 'Benefits beyond boundaries: the fishery effects of marine reserves', *Trends in Ecology and Evolution*, 18, 448-455.

Gerber, L.H., Estes, J., Crawford, T.G., Peavey, L.E. and Read, A.J. (2011) 'Managing for extinction? Conflicting conservation objectives in a large marine reserve', *Conservation Letters*, 4, 417-422.

Gerhardinger, L.C., Inui, R., Matarezi, J., Hansen, C. and Vivacqua, M. (2011a) Pirajubaé Marine Extractive Reserve – governance analysis, pp.194-205 in P.J.S. Jones, W. Qiu and E.M. De Santo (eds) *Governing Marine Protected Areas: getting the balance right – volume 2*, Technical Report to Marine & Coastal Ecosystems Branch, UNEP, Nairobi.

Gerhardinger, L.C., Godoy, E.A.S., Jones, P.J.S., Sales, G. and Ferreira, B.P. (2011b) 'Marine Protected Dramas: the flaws of the Brazilian system of marine protected areas', *Environmental Management*, 47, 630-643.

Gibbons, M., Limoges, C., Nowotony, H., Shwartzman, S., Scott, P. and Trow, M. (1994) *The New Production of Knowledge: the dynamics of science and research in contemporary societies*, Sage, London.

Giddens, A. (1984) *The Constitution of Society: outline of the theory of structuration*, University of California Press, Berkeley.

Gleason, M., McCreary, S., Miller-Henson, M., Ugoretz, J., Fox, E., Merrifield, M., McClintock, W., Serpa, P. and Hoffman, K. (2010) 'Science-based and stakeholder-driven marine protected area network planning: a successful case study from north central California', *Ocean & Coastal Management*, 53, 52–68.

Goñi, R., Adlerstein, S., Alvarez-Berastegui, D., Forcada, A., Reñones, O., Criquet, G., Polti, S., Cadiou, G., Valle, C., Lenfant, P., Bonhomme, P., Pérez-Ruzafa, A., Sánchez-Lizaso, J.L., García-Charton, J.A., Bernard, G., Stelzenmüller, V. and Planes, S. (2008) 'Spillover from six western Mediterranean marine protected areas: evidence from artisanal fisheries', *Marine Ecology Progress Series*, 366, 159–174.

Goñi, R., Hilborn, R., Díaz, D., Mallol, S. and Adlerstein, S. (2010) 'Net contribution of spillover from a marine reserve to fishery catches', *Marine Ecology Progress Series*, 400, 233–243.

Goodwin, P.P. (1998) "Hired hands' or 'local voice': understandings and experience of local participation in conservation', *Transactions of the Institute of British Geography*, 23, 481–499.

—(1999) 'The end of consensus? The impact of participatory initiatives on conceptions of conservation and the countryside in the United Kingdom', *Environment and Planning D*, 17, 383–401.

Graham, N.A.J. and McClanahan, T.R. (2013) 'The last call for marine wilderness?', *BioScience*, 63, 397–402.

Guénette, S., Lauck, T. and Clark, C. (1998) 'Marine reserves: from Beverton and Holt to the present', *Reviews in Fish Biology and Fisheries*, 8, 251–272.

Guidetti, P. (2006) 'Marine reserves reestablish lost predatory interactions and cause community changes in rocky reefs', *Ecological Applications* 16, 963–976.

—(2007) 'Potential of marine reserves to cause community-wide changes beyond their boundaries', *Conservation Biology*, 21, 540–545.

Gunderson, L.H. and Holling, C.S. (2002) *Panarchy: understanding transformations in human and natural systems*, Island Press.

Gutiérrez, N.L., Hilborn, R. and Defeo, O. (2011) 'Leadership, social capital and incentives promote successful fisheries', *Nature*, 470, 386–389.

Haaland, H. and Aas, Ø. (2010) 'Eco-tourism certification – does it make a difference? A comparison of systems from Australia, Costa Rica and Sweden', *Scandinavian Journal of Hospitality and Tourism*, 10, 375–385.

Habermas, J. (1984) *The Theory of Communicative Action; volume 1: Reason and the Rationalization of Society*, English translation by Thomas McCarthy, Beacon Press, Boston.

Halpern, B.S. (2003) 'The impact of marine reserves: do reserves work and does reserve size matter?', *Ecological Applications*, 13, 1, S117–S137.

Halpern, B.S. and Warner, R.R. (2002) 'Marine reserves have rapid and lasting effects', *Ecology Letters*, 5, 361–366.

Halpern, B.S., Gaines, S.D. and Warner, R.R. (2004) 'Confounding effects of the export of production and the displacement of fishing effort from marine reserves', *Ecological Applications*, 14, 1248–1256.

Halpern, B.S., Klein, C.J., Brown, C.J., Beger, M., Grantham, H.S., Mangubhai, S., Ruckelshaus, M., Tulloch, V.J., Watt, M., White, C. and Possingham, H.P. (2013) 'Achieving the triple bottom line in the face of inherent trade-offs amongst social equity, economic return, and conservation', *Proceedings of the National Academy of Sciences*, 110, 6229–6234.

Halpern, B.S., Longo, C., Hardy, D., McLeod, K.L., Samhouri, J.F., Katona, S.K., Kleisner, K., Lester, S.E., O'Leary, J., Ranelletti, M., Rosenberg, A.A., Scarborough, C., Selig, E.R., Best, B.D., Brumbaugh, D.R., Chapin, F.S., Crowder, L.B., Daly, K.L., Doney, S.C.,

Elfes, C., Fogarty, M.J., Gaines, S.D., Jacobsen, K.I., Karrer, L.B., Leslie, H.M., Neeley, E., Pauly, D., Polasky, S., Ris, B., St Martin, K., Stone, G.S., Sumaila, U.R. and Zeller, D. (2012) 'An index to assess the health and benefits of the global ocean', *Nature*, 488, 615–620.

Hannesson, R. (2005) 'Rights based fishing: use rights versus property rights to fish', *Reviews in Fish Biology and Fisheries*, 15, 231–241.

Hargreaves-Allen, V., Mourato, S. and Milner-Gulland, J. (2011) 'A global evaluation of coral reef management performance: are MPAs producing conservation and socio-economic improvements?', *Environmental Management*, 47, 684–700.

Harrison, H.G., Williamson, D.H., Evans, R.D., Almany, G.R., Thorrold, S.R., Russ, G.R., Feldheim, K.A., van Herwerden, L., Planes, S., Srinivasan, M., Berumen, M.L. and Jones, G.P. (2012) 'Larval export from marine reserves and the recruitment benefit for fish and fisheries', *Current Biology*, 22, R1023–R1028.

Hauser, L. and Carvalho, G.R. (2008) 'Paradigm shifts in marine fisheries genetics: ugly hypotheses slain by beautiful facts', *Fish and Fisheries*, 9, 333–362.

Hawken, P., Lovins, A.B. and Lovins, L.H. (1999) *Natural Capitalism: creating the next industrial revolution*, Little, Brown and Company, New York.

Hayes, T.M. (2006) 'Parks, people, and forest protection: an institutional assessment of the effectiveness of protected areas', *World Development*, 34, 2064–2075.

Hayes, T. and Ostrom, E. (2005) 'Conserving the world's forests: are protected areas the only way?', *Indiana Law Review*, 38, 595–617.

Hien, B.T.T. (2011) 'Ha Long Bay World Natural Heritage Area – governance analysis', pp. 136–146 in P.J.S. Jones, W. Qiu and E.M. De Santo (eds) *Governing Marine Protected Areas: getting the balance right –volume 2*, Technical Report to Marine & Coastal Ecosystems Branch, UNEP, Nairobi.

Hilborn, R. (2006) 'Faith-based fisheries', *Fisheries*, 31, 554–555.

—(2010) 'Apocalypse forestalled: why all the world's fisheries aren't collapsing', *The Science Chronicles*, November 2010, 5–9.

Hilborn, R., Micheli, F. and De Leo, G.A. (2006) 'Integrating marine protected areas with catch regulation', *Canadian Journal of Fisheries and Aquatic Sciences*, 63, 642–649.

Hilborn, R., Orensanz, J.M. and Parma, A.M. (2005) 'Institutions, incentives and the future of fisheries', *Philosophical Transactions of the Royal Society B*, 360, 47–57.

Hilborn, R., Punt, A.E. and Orensanz, J. (2004) 'Beyond band-aids in fisheries management: fixing world fisheries', *Bulletin of Marine Science*, 74, 493–507.

Holles, S., Simpson, S.D., Radford, A.N., Berten, E. and Lecchini, D. (2013) 'Boat noise disrupts orientation behaviour in a coral reef fish', *Marine Ecology Progress Series*, 485, 295–300.

Holling, C.S. (1973) 'Resilience and stability of ecological systems', *Annual Review of Ecology and Systematics*, 4, 1–23.

—(ed.) (1978) *Adaptive environmental assessment and management*, Wiley, New York.

Holmes, B. (1997) 'Dreams as big as an ocean', *New Scientist*, 155, 20.

Homewood, C., Kristjanson, P. and Trench, P.C. (eds) (2009) *Staying Maasai?: livelihoods, conservation and development in East African Rangelands*, Springer, New York.

Howarth, L.M., Roberts, C.M., Thurstan, R.H. and Stewart, B.D. (2014) 'The unintended consequences of simplifying the sea: making the case for complexity', *Fish and Fisheries*, DOI:10.1111/faf.12041.

Hughes, T.P. (1994) 'Catastrophes, phase shifts, and large-scale degradation of a Caribbean coral reef', *Science*, 265, 1547–1551.

Hughes, T.P., Baird, A.H., Bellwood, D.R., Card, M., Connolly, S.R., Folke, C., Grosberg, R., Hoegh-Guldberg, O., Jackson, J.B.C., Kleypas, J., Lough, J.M., Marshall, P., Nyström, M., Palumbi, S.R., Pandolfi, J.M., Rosen, B. and Roughgarden, J. (2003) 'Climate change, human impacts, and the resilience of coral reefs', *Science*, 301, 929-933.

Hughes, T.P., Bellwood, D.R., Folke, C., Steneck, R.S. and Wilson, J. (2005) 'New paradigms for supporting the resilience of marine ecosystems', *Trends in Ecology & Evolution*, 20, 7, 380-386.

Hutton, J., Adams, W.M. and Murombedzi, J.C. (2005) 'Back to the barriers? Changing narratives in biodiversity conservation', *Forum for Development Studies* 2, 341-370.

IPSO (2011) *Implementing the Global State of the Oceans Report*. International Programme on the State of the Ocean, Oxford, UK.

IUCN/UNEP-WCMC (2012) *The World Database on Protected Areas (WDPA): January 2011*, UNEP-WCMC, Cambridge, UK.

IUCN-WCPA (IUCN World Commission on Protected Areas, 2008). *Establishing Marine Protected Area Networks – Making It Happen*, IUCN-WCPA, National Oceanic and Atmospheric Administration and The Nature Conservancy, Washington, D.C..

Imperial, M.T. (1999) 'Institutional analysis and ecosystem-based management: the institutional analysis and development framework', *Environmental Management*, 24, 449-465.

Jackson, J.B.C., Kirby, M.X., Berger, W.H., Bjorndal, K.A., Botsford, L.W., Bourque, B.J., Bradbury, R.H., Cooke, R., Erlandson, J., Estes, J.A., Hughes, T.P., Kidwell, S., Lange, C.B., Lenihan, H.S., Pandolfi, J.M., Peterson, C.H., Steneck, R.S., Tegner, M.J. and Warner, R.R. (2001) 'Historical overfishing and the recent collapse of coastal ecosystems', *Science*, 293, 629-638.

Jacquet, J., Hocevar, J., Lai, S., Majluf, P., Pelletier, N., Pitcher, T., Sala, E., Sumaila, R. and Pauly, D. (2009) 'Conserving wild fish in a sea of market-based efforts', *Oryx*, 44, 45-56.

Jennings, S. and Kaiser, M.J. (1998) 'The effects of fishing on marine ecosystems', *Advances in Marine Biology*, 34, 203-352.

Jennings, S., Kaiser, M.J. and Reynolds, J.D. (2001) *Marine Fisheries Ecology*, Blackwell, Oxford, UK.

Jentoft, S. (2004) 'Institutions in fisheries: what they are, what they do, and how they change', *Marine Policy*, 28, 137-149.

Jentoft, S. and Chuenpagdee, R. (2009) 'Fisheries and coastal governance as a wicked problem', *Marine Policy*, 33, 553-560.

Jentoft, S., Chuenpagdee, R. and Pascual-Fernández, J.J. (2011) 'What are MPAs for: on goal formation and displacement', *Ocean & Coastal Management*, 54, 75-83.

Jiao, Y. (2009) 'Regime shift in marine ecosystems and implications for fisheries management, a review', *Reviews in Fish Biology and Fisheries*, 19, 177-191.

Johannes, R.E. (1978) 'Traditional marine conservation methods in Oceania and their demise', *Annual Review of Ecology and Systematics*, 9, 349-364.

—(2002) 'The renaissance of community-based marine resource management in Oceania', *Annual Review of Ecology and Systematics*, 33, 317-340.

Johnston, R.J., Gegory,. D., Pratt, G. and Watts, M. (2000) *The Dictionary of Human Geography (Fourth Edition)*, Wiley-Blackwell, Oxford.

Jones, P.J.S. (1994) 'A review and analysis of the objectives of marine nature reserves', *Ocean & Coastal Management*, 24, 149-178.

—(2001) 'Marine protected area strategies: issues, divergences and the search for middle ground', *Reviews in Fish Biology and Fisheries*, 11, 197-216.

—(2006) 'Collective action problems posed by no-take zones', *Marine Policy*, 30, 143–156.

—(2007) 'Point of View – Arguments for conventional fisheries management and against no-take marine protected areas: only half of the story?', *Reviews in Fish Biology and Fisheries* 17, 31–43.

—(2008) 'Fishing industry and related perspectives on the issues raised by no-take marine protected area proposals', *Marine Policy*, 32, 749–758.

—(2009) 'Equity, justice and power issues raised by no-take marine protected area proposals', *Marine Policy*, 33, 759–765.

—(2011) 'The Wash & North Norfolk Coast European Marine Site – governance analysis', pp. 40–59 in P.J.S. Jones, W. Qiu and E.M. De Santo (eds) *Governing Marine Protected Areas: getting the balance right –volume 2*, Technical Report to Marine & Coastal Ecosystems Branch, UNEP, Nairobi.

—(2012) 'Marine protected areas in the UK: challenges in combining top-down and bottom-up approaches to governance', *Environmental Conservation*, 39, 248–258.

—(2013a) 'Governing protected areas to fulfil biodiversity conservation obligations: from Habermasian ideals to a more instrumental reality', *Environment, Development and Sustainability*, 15, 39–50.

—(2013b) 'A governance analysis of the Galápagos Marine Reserve', *Marine Policy*, 41, 65–71.

Jones, P.J.S. and Burgess, J. (2005) 'Building partnership capacity for the collaborative management of marine protected areas in the UK: a preliminary analysis', *Journal of Environmental Management*, 77, 227–243.

Jones, P.J.S. and Carter, A. (2009) 'Crossing the divide: the challenges of designing an ecologically coherent and representative network of MPAs for the UK', *Marine Policy*, 33, 5, 737–743.

Jones, P.J.S., Qiu, W. and De Santo, E.M. (2011) *Governing MPAs: getting the balance right*, Technical Report to UNEP, http://www.mpag.info, accessed 12 March 2013.

Jones, P.J.S., De Santo, E.M, Qiu, W. and Vestergaard O. (2013a) 'Introduction: an empirical framework for deconstructing the realities of governing marine protected areas', *Marine Policy*, 41, 1–4.

Jones, P.J.S., Qiu, W. and De Santo, E.M. (2013b) 'Governing marine protected areas: social-ecological resilience through institutional diversity', *Marine Policy*, 41, 5–13.

Kaiser, M.J. (2005) 'Are marine protected areas a red herring or fisheries panacea?', *Canadian Journal of Fisheries and Aquatic Sciences*, 62, 1194–1199.

Kareiva, P. (2003) 'Marine reserves: the best option for our oceans?', *Frontiers in Ecology and the Environment*, 1, 501–502.

Kaufmann, D., Kraay, A. and Mastruzzi, M. (2009) *Governance Matters VIII: aggregate and individual governance indicators 1996–2008*, World Bank Policy Research Working Paper No. 4978, World Bank, Washington DC.

Kelleher, G. (1999) *Guidelines for Marine Protected Areas*, IUCN, Gland/Cambridge.

Kelleher, G. and Kenchington, R. (1991) *Guidelines for Establishing Marine Protected Areas*, IUCN, Gland/Cambridge.

Kelleher, G. and Kenchington, R.A. (1982) 'Australia's Great Barrier Reef Marine Park: making development compatible with conservation', Ambio, 11, 5, 262–267.

Kelleher, G., Bleakley, C. and Wells, C. (1995) *Priority Areas for a Global Representative System of Marine Protected Areas*, Great Barrier Reef Marine Park Authority, The World Bank and the World Conservation Union (IUCN), Environment Department, World Bank, Washington D.C.

Kellner, J.B., Tetreault, I., Gaines, S.D. and Nisbet, R.M. (2007) 'Fishing the line near marine reserves in single and multispecies fisheries', *Ecological Applications*, 17, 1039–1054.

Kenchington, R.A. (1990) *Managing Marine Environments*, Taylor and Francis, New York.

——(1993) 'Tourism in coastal and marine environments – a recreational perspective', *Ocean & Coastal Management* 19, 1–16.

Kenchington, R.A. and Agardy, T. (1990) 'Achieving marine conservation through biosphere reserve planning and management', *Environmental Conservation*, 17, 39–44.

Kinlan, B.P. and Gaines, S.D. (2003) 'Propagule dispersal in marine and terrestrial environments: a community perspective', *Ecology*, 84, 2007–2020.

Kjær, A.M. (2004) *Governance*, Polity Press, Cambridge.

Knowlton, N. (1992) 'Thresholds and multiple stable states in coral reef community dynamics', *American Zoologist*, 32, 674–682.

Kooiman, J., Bavinck, M., Chuenpagdee, R., Mahon, R. and Pullin, R. (2008) 'Interactive governance and governability: an introduction', *The Journal of Transdisciplinary Environmental Studies*, 7, 1–11.

Kronenberg, J. and Hubacek, K. (2013) 'Could payments for ecosystems services create an "ecosystems services curse"', *Ecology and Society*, 18, 1, article 10.

Lane, M.B. and Corbett, T. (2005) 'The tyranny of localism: indigenous participation in community-based environmental management', *Journal of Environmental Policy & Planning*, 7, 141–159.

Larkin, P.A. (1977) 'An epitaph to the concept of maximum sustainable yield', *Transactions of the American Fisheries Society*, 106, 1–11.

Lau, W.W.Y. (2012) 'Beyond carbon: conceptualizing payments for ecosystem services in blue forests on carbon and other marine and coastal ecosystem services', *Ocean & Coastal Management*, 83, 5–14.

Lauck, T., Clark, C.W., Mangel, M. and Munro, G.R. (1998) 'Implementing the precautionary principle in fisheries management through marine reserves', *Ecological Applications*, 8, S72–S78.

Law, R. (2000) 'Fishing, selection and phenotypic evolution', *ICES Journal of Marine Science*, 57, 659–668.

Lester, S.E. and Halpern, B.S. (2008) 'Biological responses in marine no-take reserves versus partially protected areas', *Marine Ecology Progress Series* 367, 49–56.

Lester, S.E., Halpern, B.S., Grorud-Colvert, K., Lubchenco, J., Ruttenberg, B.I., Gaines, S.D., Airamé, S. and Warner, R.R. (2009) 'Biological effects within no-take marine reserves: a global synthesis', *Marine Ecology Progress Series*, 384, 33–46.

Ling, S.D. and Johnson, C.R. (2012) 'Marine reserves reduce risk of climate-driven phase shift by reinstating size- and habitat-specific trophic interactions', *Ecological Applications*, 22, 1232–1245.

LoBue, C. and Udelhoven, J. (2013) 'Private ownership of underwater lands in Great South Bay, New York: a case study in degradation, restoration and protection', *Marine Policy*, 41, 103–109.

Longhurst, A. (2002) 'Murphy's law revisited: longevity as a factor in recruitment to fish populations', *Fisheries Research*, 56, 125–131.

Low, B., Ostrom, E., Simon, C. and Wilson, J. (2003) 'Redundancy in social and ecological systems', pp. 83–114 in F. Berkes, J. Colding and C. Folke (eds) *Navigating Social-ecological*

Systems: building resilience for complexity and change, Cambridge University Press, Cambridge, UK.

Lubchenco, J., Palumbi, S.R., Gaines, S.D. and Andelman, S. (2003) 'Plugging a hole in the ocean: the emerging science of marine reserves', *Ecological Applications*, 13, 1 (Supplement), S3-S7.

Lucas, E.Y. and Kirit, R. (2009) 'Fisheries-marine protected area-tourism interactions in Moalboal, Cebu, Philippines', *Coastal Management*, 37, 480-490.

Ludwig, D., Hilborn, R. and Walters, C. (1993) 'Uncertainty, resource exploitation, and conservation: lessons from history', *Science*, 260, 17-36.

Lynam, C.P., Gibbons, M.J., Axelsen, B.E., Sparks, C.A.J., Coetzee, J., Hewwood, B.G. and Brierley, A.S. (2006) 'Jellyfish overtake fish in heavily fished ecosystem', *Current Biology*, 16, 13, R492-R493.

Lyster, S. (1985) *International Wildlife Law: an analysis of international treatise concerned with the conservation of wildlife*, Cambridge University Press, Cambridge, UK.

MCBI (Marine Conservation Biology Institute, 1998) *Troubled Waters: a call for action*, http://www.mcbi.org/AboutUs/TroubledWaters.pdf, accessed 15 January 2012.

McCann, K., Hastings, A. and Huxel, G.R. (1998) 'Weak trophic interactions and the balance of nature', *Nature*, 395, 794-798.

McCarthy, J. and Prudham, S. (2004) 'Neoliberal nature and the nature of neoliberalism', *Geoforum*, 35, 275-283.

McCay, B.J. (2002) 'Emergence of institutions for the commons: contexts, situations and events', pp. 361-402 in E. Ostrom, T. Dietz, N. Dolšak, P.C. Stern, S. Sonich and E.U. Weber (eds) *The Drama of the Commons,*. National Academy Press, Washington DC.

McCay, B., Powell, E. and Bochenek, E. (2008) 'Climate change, clam die-offs and industry response to an ITQ fishery', paper presented at the 12th Biennial Conference of the International Association for the Study of the Commons - *Governing Shared Resources: connecting local experience to global challenges*, July 2008, University of Gloucestershire, UK.

McClanahan, T. (1999) 'Is there a future for coral reef parks in poor tropical countries?', *Coral Reefs*, 18, 321-325.

——(2004) 'The limits to beyond boundaries', *Aquatic Conservation*, 14, 1-4.

McClanahan, T.R., Ateweberhan, M., Graham, N.A.J., Wilson, S.K., Sebastián, R.C., Guillaume, M.M.M. and Bruggemann, J.H. (2007) 'Western Indian Ocean coral communities: bleaching responses and susceptibility to extinction', *Marine Ecology Progress Series*, 337, 1-13.

McClanahan, T.R., Graham, N.A.J., Calnan, J.M. and MacNeil, M.A. (2007) 'Toward pristine biomass: reef fish recovery in coral reef marine protected areas in Kenya', *Ecological Applications*, 17, 1055-1067.

McCook, L.J., Ayling, T., Cappo, M., Choat, J.H., Evans, R.D., De Freitas, D.M., Heupel, M., Hughes, T.P., Jones, G.P., Mapstone, B., Marsh, H., Mills, M., Molloy, F.J., Pitcher, C.R., Pressey, R.L., Russ, G.R., Sutton, S., Sweatman, H., Tobin, R., Wachenfield, D.R. and Williamson, D.H. (2010) 'Adapative management of the Great Barrier Reef: globally significant demonstration of the benefits of networks of marine reserves', *Proceedings of the National Academy of Sciences*, 107, 18278-18285.

McCrea-Strub, A., Zeller, D., Sumaila, U.R., Nelson, J., Balmford, A. and Pauly, D. (2011) 'Understanding the cost of establishing marine protected areas', *Marine Policy*, 35, 1-9.

McGinnis, M. (2000) *Polycentric Governance and Development*, University Michigan Press, Ann Arbor, MI.

McIntyre, A.D. (1992) 'Introduction: a perspective on marine conservation', *Proceedings of the Royal Society of Edinburgh*, Section B, Biological Sciences, 100, 1–2.

McLeod, K.L., Lubchenco, J., Palumbi, S.R. and Rosenberg, A.A. (2005) *Scientific Consensus Statement on Marine Ecosystem-Based Management*, Communication Partnership for Science and the Sea.

McManus, J.W. (1994) 'The Spratly Islands: a Marine Park?', *Ambio*, 23, 3, 181–186.

Macedo, H.S., Vivacqua, M., Rodrigues, H.C.L. and Gerhardinger, L.C. (2013) 'Governing wide coastal-marine protected territories: a governance analysis of the Baleia Franca Environmental Protection Area in South Brazil', *Marine Policy*, 41, 118–125.

Macintosh, A., Bonyhady, T. and Wilkinson, D. (2010) 'Dealing with interests displaced by marine protected areas: a case study on the Great Barrier Reef Marine Park structural adjustment package', *Ocean & Coastal Management*, 53, 581–588.

Mackelworth, P., Holcer, D. and Fortuna, C.M. (2013) 'Unbalanced governance: the Cres-Lošinj Special Marine Reserve, a missed conservation opportunity', *Marine Policy*, 41, 126–133.

Maina, J., Venus, V., McClanahan, T. and Ateweberhan, M. (2008) 'Modelling susceptibility of coral reefs to environmental stress using remote sending data and GIS models', *Ecological Modelling*, 212, 180–199.

Marinesque, S., Kaplan, D.M. and Rodwell, L.D. (2012) 'Global implementation of marine protected areas: is the developing world being left behind?', *Marine Policy*, 36, 727–737.

Marteinsdottir, G. and Steinarsson, A. (1998) 'Maternal influence on the size and viability of cod (*Gadus morhua* L.) eggs and larvae', *Journal of Fish Biology*, 52, 1241–1258.

Mascia, M.B. and Claus, C.A. (2009) 'A property rights approach to understanding human displacement from protected areas: the case of marine protected areas', *Conservation Biology*, 23, 16–23..

Mascia, M.B., Claus, C.A. and Naidoo, R. (2010) 'Impacts of marine protected areas on fishing communities', *Conservation Biology*, 24, 1424–1429.

Micheli, F. and Halpern, B.S. (2005) 'Low functional redundancy in coastal marine assemblages', *Ecology Letters*, 8, 391–400.

Micheli, F., Saenz-Arroyo, A., Greenley, A., Vazquez, L., Espinoza Montes, J.A., Rossetto, M. and De Leo, G.A. (2012) 'Evidence that marine reserves enhance resilience to climatic impacts', *PLoS ONE*, 7, 7, e40832.

Micheli, F., Halpern, B.S., Botsford, L.W. and Warner, R.R. (2004) 'Trajectories and correlates of community change in no-take marine reserves', *Ecological Applications*, 14, 1709–1723.

Miller, M.L. and Kirk, J. (1992) 'Marine environmental ethics', *Ocean & Coastal Management*, 17, 237–251.

Moberg, F. and Folke, C. (1999) 'Ecological goods and services of coral reef ecosystems', *Ecological Economics*, 29, 215–233.

Mol, A.P.J. and Sonnenfeld, D.A. (2000) 'Modernisation around the world: an introduction', *Environmental Politics*, 9, 3–16.

Moland, E., Olsen, E.M., Knutsen, H., Garrigou, P., Espeland, S.H., Kleiven, A.R., André, C. and Knutsen, J.A. (2013) 'Lobster and cod benefit from small-scale northern marine protected areas: inference from an empirical before-after control-impact study', *Proceedings of the Royal Society B*, 280, article 20122679.

Molloy, P.P., McLean, I.B. and Côté, I.M. (2009) 'Effects of marine reserve age on fish populations: a global meta-analysis', *Journal of Applied Ecology* 46, 743–751.

Monte-Luna, P.D., Lluch-Belda, D., Serviere-Zaragoza, E., Carmona, R., Reyes-Bonilla, H., Aurioles-Gamboa, D., Castro-Aguirre, J.L., Guzmán del Próo, S.A., Trujillo-Millán, O. and Brook, B.W. (2007) 'Marine extinctions revisited', *Fish and Fisheries*, 8, 107-122.

Moore, E. (2011) 'National Marine Sanctuary System, USA – governance analysis', pp. 60-64 in P.J.S. Jones, W. Qiu and E.M. De Santo (eds) *Governing Marine Protected Areas: getting the balance right –volume 2*, Technical Report to Marine & Coastal Ecosystems Branch, UNEP, Nairobi.

Mora, C., Andréfouët, S., Costello, M.J., Rollo, A., Veron, J., Gaston, K.J. and Myers, R.A. (2006) 'Coral reefs and the global network of marine protected areas', *Science*, 312, 1750-1751.

Morita, K., Yamamoto, S., Takashima, Y., Matsuishi, T., Kanno, Y. and Nishimura, K. (1999) 'Effect of maternal growth history on egg number and size in wild white-spotted char (*Salvelinus luecomaenis*)', *Canadian Journal of Fisheries and Aquatic Sciences*, 56, 1585-1589.

Mosquera, I., Côté, I.M., Jennings, S. and Reynolds, J.D. (2000) 'Conservation benefits of marine reserves for fish populations', *Animal Conservation*, 4, 321-332.

Mumby, P.J. and Harborne, A.R. (2010) 'Marine reserves enhance the recovery of corals on Caribbean reefs', *PLoS One*, 5, 1, e8657.

Mumby, P.J., Hastings, A. and Edwards, H.J. (2007) 'Thresholds and the resilience of Caribbean coral reefs', *Nature*, 450, 98-101.

Mumby, P.R., Dahlgren, C.P., Harborne, A.R., Kappel, C.V., Micheli, F., Brumbaugh, D.R., Holmes, K., Mendes, J.M., Broad, K., Sanchirico, J.N., Buch, K., Box, S., Stoffle, R.W. and Gill, A.B. (2006) 'Fishing, trophic cascades, and the process of grazing on coral reefs', *Science*, 311, 98-101.

Murawski, S.A., Wigley, S.E., Fogarty, M.J., Rago, P.J. and Mountain, D.G. (2005) 'Effort distribution and catch patterns adjacent to temperate MPAs', *ICES Journal of Marine Science* 62, 1150-1167.

Murphree, M.W. (1994) 'The role of institutions in community-based conservation', pp. 403-427 in D. Western, R.M. Wright and S.C. Strum (eds) *Natural Connections: perspectives in community-based conservation*, Island Press, Washington DC.

Murray, S.N., Ambrose, R.F., Bohnsack, J.A., Botsford, L.W., Carr, M.H., Davis, G.E., Dayton, P.K., Gotshall, D., Gunderson, D.R., Hixon, M.A., Lubchenco, J., Mangel, M., MacCall, A., McArdle, D.A., Ogden, J.C., Roughgarden, J., Starr, R.M., Tegner, M.J. and Yoklavich, M.M. (1999) 'No-take reserve networks: sustaining fishery populations and marine ecosystems', *Fisheries*, 24, 11, 11-25.

Myers, R.A., Baum, J.K., Shepherd, T.D., Powers, S.P. and Peterson, C.H. (2007) 'Cascading effects of the loss of apex predatory sharks from a coastal ocean', *Science*, 315, 1846-1850.

Myers, R.A. and Worm, B. (2003) 'Rapid worldwide depletion of predatory fish communities', *Nature*, 423, 280-283.

NCEAS (National Centre for Ecological Analysis and Synthesis, 2001) 'Scientific consensus statement on marine reserves and protected areas', *Journal of International Wildlife Law & Policy*, 4, 1, 87-90, http://www.nceas.ucsb.edu/Consensus/consensus.pdf, accessed 16 January 2012.

NRC (National Research Council, 2001) *Marine Protected Areas: tools for sustaining ocean ecosystems*, National Academy Press, Washington, DC.

Newig, J. and Fritsch, O. (2009) 'Environmental governance: participatory, multi-level – and effective?' *Environmental Policy and Governance*, 19, 197-214.

Nichols, K. (1999) 'Coming to terms with "integrated coastal management": problems of meaning and method in a new arena of resource regulation', *Professional Geographer*, 51, 388–399.

Nordlund, L.M., Kloiber, U., Carter, E. and Riedmiller, S. (2013) 'Chumbe Island Coral Park – governance analysis', *Marine Policy*, 41, 110–117.

Norse, E. (2010) 'Ecosystem-based spatial planning and management of marine fisheries: why and how?', *Bulletin of Marine Science*, 86, 179–195.

Nowotony, H., Scott, P. and Gibbons, M. (2001) *Re-thinking Science: knowledge and the public in the age of uncertainty*, Polity Press, Cambridge.

Nyström, M., Folke, C. and Moberg, F. (2000) 'Coral reef disturbance and resilience in a human-dominated environment', *Trends in Ecology & Evolution*, 15, 413–417.

Olsen, E., Kleiven, A.R., Skjoldal, H.R. and von Quillfeldt, C.H. (2011) 'Place-based management at different scales', *Journal of Coastal Conservation* 15, 257–269.

Olsson, P., Folke, C. and Hughes, T.P. (2008) 'Navigating the transition to ecosystem-based management of the Great Barrier Reef, Australia', *Proceedings of the National Academy of Sciences*, 105, 9489–9494.

Oracion, E.G., Miller, M.L. and Christie, P. (2005) 'Marine protected areas for whom? Fisheries, tourism and solidarity in a Philippine community', *Ocean & Coastal Management*, 48, 393–410.

Ostrom, E. (1990) *Governing the Commons: the evolution of institutions for collective action*, Cambridge University Press, Cambridge, UK.

——(1995) *Understanding Institutional Diversity*, Princeton University Press, Princeton, New Jersey.

——(1996) 'Crossing the great divide: coproduction, synergy, and development', *World Development*, 2, 1073–1087.

——(1998) 'A behavioural approach to the rational choice theory of collective action', *American Political Science Review*, 92, 1–22.

——(1999) 'Coping with tragedies of the commons', *Annual Review of Political Science*, 2, 493–535.

——(2007) 'Going beyond panaceas', *Proceedings of the National Academy of Sciences*, 104, 15176–15178.

——(2009) 'A general framework for analysing sustainability of social-ecological systems', *Science*, 325, 419–422.

Ostrom, E. and Cox, M. (2010) 'Moving beyond panaceas: a multi-tiered diagnostic approach for social-ecological analysis', *Environmental Conservation*, 37, 451–463.

Oxhorn, P. (2004) 'Unraveling the puzzle of decentralization', pp. 3–32 in P. Oxhorn, J.S. Tulchin and A.D. Selee (eds) *Decentralization, Democratic Governance, and Civil Society in Comparative Perspective: Africa, Asia, and Latin America*, John Hopkins University Press, Baltimore.

Pauly, D. (1995) 'Anecdotes and the shifting baseline syndrome of fisheries', *Trends in Ecology and Evolution*, 10, 430.

——(2009a) 'Aquacalypse now: the end of fish', *The New Republic*, 240, 24–27.

——(2009b) 'Beyond duplicity and ignorance in global fisheries', *Scientia Marina*, 73, 215–224.

Pauly, D. and Christensen, V. (1995) 'Primary production required to sustain global fisheries', *Nature*, 1374, 255–257.

Pauly, D. and Froese, R. (2012) 'Comments on FAO's State of Fisheries and Aquaculture, or "SOFIA2010"', *Marine Policy*, 36, 745–752.

Pauly, D., Watson, R. and Alder, J. (2005) 'Global trends in world fisheries: impacts on marine ecosystems and food security', *Philosophical Transactions of the Royal Society B*, 360, 1453, 5-12.

Pauly, D., Christensen, V., Dalsgaard, J., Froese, R. and Torres, F. (1998) 'Fishing down marine food webs', *Science*, 279, 860-863.

Pauly, D., Christensen, V., Guénette, S., Pitcher, T.J., Sumaila, U.R., Walters, C.J., Watson, R. and Zeller, D. (2000) 'Towards sustainability in world fisheries', *Nature*, 418, 689-695.

——(2002) 'Towards sustainability in world fisheries', *Nature* 418, 689-695.

Pearsall, S.H. (1984) '*In absentia* benefits of nature preserves: a review', *Environmental Conservation*, 11, 3-10.

Pennington, M. and Rydin, Y. (2000) 'Researching social capital in local environmental policy contexts', *Policy & Politics*, 28, 233-249.

Perez de Oliveira, L. (2013) 'A governance analysis of Os Miñarzos Marine Reserve of Fishing Interest', *Marine Policy*, 41, 95-102.

Peterson, A.M. and Stead, S.M. (2010) 'Rule breaking and livelihood options in marine protected areas', *Environmental Conservation*, 38, 342-352.

Phillips, A. (2003) 'Turning ideas on their head: the new paradigm for protected areas', *George Wright Forum*, 20, 2, 8-32.

Pichegru, L., Grémillet, D., Crawford, R.J.M. and Ryan, P.G. (2010) 'Marine no-take zone rapidly benefits endangered penguin', *Conservation Letters*, 6, 498-501.

Pikitch, E.K., Santora, C., Babcock, E.A., Bakun, A., Bonfil, R., Conover, D.O., Dayton, P., Doukakis, P., Fluharty, D., Heneman, B., Houde, E.D., Link, J., Livingston, P.A., Mangel, M., McAllister, M.K., Pope, J. and Sainsbury, K.J. (2004) 'Ecosystem-based fishery management', *Nature*, 305, 346-347.

Pinkerton, E.W. (1992) 'Translating legal rights into management practice: overcoming barriers to the exercise of co-management', *Human Organization*, 51, 330-341.

Platteau, J.P. (2004) 'Monitoring elite capture in community-driven development', *Development and Change*, 35, 223-246.

Polidoro, B.A., Livingstone, S.R., Carpenter, K.E., Hutchinson, B., Mast, R.B., Pilcher, N., Sadovy de Mitcheson, Y. and Valenti, S. (2008) 'Status of the world's marine species' in J.-C. Vié, C. Hilton-Taylor and S.N. Stuart (eds) *The 2008 Review of The IUCN Red List of Threatened Species*, IUCN, Gland, Switzerland.

Polis, G.A. (1998) 'Stability is woven by complex webs', *Nature*, 395, 744-745.

Polunin, N.V.C. (1984) 'Do traditional marine "reserves" conserve? A view of Indonesian and New Guinean evidence', *Senri Ethnological Studies*, 17, 267-283.

Pomeroy, R.S., Parks, J.E. and Watson, L.M. (2004) *How is your MPA doing? A Guidebook of Natural and Social Indicators for Evaluating Marine Protected Area Management Effectiveness*, IUCN, Gland, Switzerland.

Pratchett, M. (2005) 'Dynamics of an outbreak population of *Acanthaster planci* at Lizard Island, northern Great Barrier Reef (1995-1999)', *Coral Reefs*, 24, 453-462.

Prideaux, M., Emmet, J. and Horstman, M. (1998) 'Sustainable use or multiple abuse?', *Habitat Australia*, 26, 2, 13-20.

Qiu, W. (2013) 'The Sanya Coral Reef National Marine Nature Reserve, China: a governance analysis', *Marine Policy*, 41, 50-56.

Qiu, W and Jones, P.J.S. (2013) 'The emerging policy landscape for marine spatial planning in Europe', *Marine Policy*, 39, 182-190.

Raik, D.B., Wilson, A.L. and Decker, D.J. (2008) 'Power in natural resources management: an application of theory', *Society and Natural Resources*, 21, 729-739.

Ravetz, J.R. (1999) 'What is post-normal science?', *Futures*, 31, 647–653.

Ray, G.C. (1962) 'Inshore Marine Conservation', pp. 79–87 in A.B. Adams (ed.) *First World Conference on National Parks, Seattle, Washington*, National Park Service, Washington DC.

——(1976) 'Critical marine habitats', pp. 15–59 in K.K. Sentā (ed.) *Proceedings of an International Conference on Marine Parks and Reserves*, 12–14 May, 1975, Tokyo, Japan, IUCN, Gland, Switzerland.

——(1996) 'Coastal-marine discontinuities and synergisms: implications for biodiversity and conservation', *Biodiversity Conservation*, 5, 1095–1108.

——(1999) 'Coastal-marine protected areas: agonies of choice', *Aquatic Conservation: Marine and Freshwater Ecosystems*, 9, 6, 607–614.

Redford, K.H. and Adams, W.M. (2009) 'Payment for ecosystem services and the challenge of saving nature', *Conservation Biology*, 23, 785–787.

Rees, S.E., Austen, M.C., Attrill, A.J. and Rodwell, L.D. (2012) 'Incorporating indirect ecosystem services into marine protected area planning and management', *International Journal of Biodiversity Science, Ecosystem Services & Management*, 8, 273–285.

Reid, P.C., Battle, E.J.V., Batten, S.D. and Brander, K.M. (2000) 'Impacts of fisheries on plankton community structure', *ICES Journal of Marine Science*, 57, 495–502.

Rice, J. (2011) 'Managing fisheries well: delivering the promises of an ecosystem approach', *Fish and Fisheries*, 12, 209–231.

Rice, J.C. and Garcia, S.M. (2011) 'Fisheries, food security, climate change, and biodiversity: characteristics of the sector and perspectives on emerging issues', *ICES Journal of Marine Science*, 68, 1343–1353.

Rice, J. and Houston, K. (2011) 'Representativity and networks of marine protected areas', *Aquatic Conservation: Marine and Freshwater Ecosystems*, 21, 649–657.

Rice, J., Mokness, E., Attwood, C., Brown, S.K., Dahle, G., Gjerde, K.M., Grefsrud, S., Kenchington, R., Kleiven, A.R., McConney, P., Ngoile, M.A.K., Næsje, T.F., Olsen, E., Olsen, E.M., Sanders, J., Sharma, C., Vestergaard, O. and Westlund, L. (2012) 'The role of MPAs in reconciling fisheries management with conservation of biological diversity', *Ocean & Coastal Management*, 69, 217–230.

Rip, A. (2006) 'A co-evolutionary approach to reflexive governance and its ironies', pp. 82–100 in J.P. Voss, D. Bauknecht and R. Kemp (eds) *Reflexive Governance for Sustainable Development*, Edward Elgar, Cheltenham.

Risbey, J.S. (2008) 'The new climate discourse: alarmist or alarming?', *Global Environmental Change*, 16, 26–37.

Roberts, C.M. (1995) 'Effects of fishing on the ecosystem structure of coral reefs', *Conservation Biology*, 9, 5, 988–995.

——(1997a) 'Connectivity and management of Caribbean coral reefs', *Science*, 278, 1454–1457.

——(1997b) 'Ecological advice for the global fisheries crisis', *Trends in Ecology and Evolution*, 2, 35–38.

——(1998) 'Sources, sinks, and the design of marine reserve networks', *Fisheries*, 23, 7, 16–19.

——(2000) 'Selecting marine reserve locations: optimality versus opportunism', *Bulletin of Marine Science*, 66, 3, 581–592.

——(2007) *The Unnatural History of the Sea: the past and future of humanity and fishing*. Gaia, London (published by Island Press in US).

——(2012a) *The Ocean of Life: the fate of man and the sea*, Viking Press, New York.

——(2012b) 'Marine ecology: reserves do have a role in fisheries', *Current Biology*, 22, R444–R446.

Roberts, C.M. and Hawkins, J.P. (1999) 'Extinction risk in the sea', *Trends in Research in Ecology and Evolution*, 14, 241-246.

Roberts, C.M., Bohnsack, J.A., Gell, F., Hawkins, J.P. and Goodridge, R. (2001) 'Effects of marine reserves on adjacent fisheries, *Science*, 294, 1920-1923.

Roberts, C.M., Hawkins, J.P. and Gell, F.R. (2005) 'The role of marine reserves in achieving sustainable fisheries', *Philosophical Transactions of the Royal Society B*, 360, 123-132.

Roberts, T. and Jones, P.J.S. (2009) 'Shellfishing, eider ducks and nature conservation on the Wash: questions raised by a fractured partnership', *Society & Natural Resources*, 22, 538-553.

—(2013) 'North East Kent European Marine Site: overcoming barriers to conservation through community engagement', *Marine Policy*, 41, 33-40.

Roff, J. and Zacharias, M. (2011) *Marine Conservation Ecology*, Earthscan, London.

Roman, G.S.J., Dearden, P. and Rollins, R. (2007) 'Application of zoning and "limits of acceptable change" to manage snorkelling tourism', *Environmental Management*, 39, 819-830.

Rondinelli, D. (2000) 'What is decentralization?', pp. 2-5 in J. Litvack and J. Seddon (eds) *Decentralization Briefing Notes*, World Bank Institute in collaboration with PREM network, Washington DC.

Rudd, M.A., Tupper, M.H., Folmer, H. and van Kooten, G.C. (2003) 'Policy analysis for tropical marine reserves: challenges and directions', *Fish and Fisheries*, 4, 65-85.

Ruhl, J.B. (2012) 'Panarchy and the law', *Ecology and Society*, 17, 3, article 31.

Russ, G.R. and Alcala, A.C. (2011) 'Enhanced biodiversity beyond marine reserve boundaries: The cup spillith over', *Ecological Applications*, 21, 241-250.

Russ, G.R., Alcala, A.C., Maypa, A.P., Calumpong, H.P. and White, A.T. (2004) 'Marine reserve benefits local fisheries', *Ecological Applications*, 14, 597-606.

Russ, G.R., Stockwell, B. and Alcala, A.C. (2005) 'Inferring versus measuring rates of recovery in no-take marine reserves', *Marine Ecology Progress Series*, 292, 1-12.

Rydin, Y. (2006) Institutions and networks: the search for conceptual research tools, pp. 15-33 in Y. Rydin and E. Falleth (eds) *Networks and Institutions in Natural Resource Management*, Edward Elgar Publishing, Cheltenham, UK.

Rydin, Y. and Holman, N. (2004) 'Re-evaluating the contribution of social capital in achieving sustainable development', *Local Environment*, 9, 117-133.

Rydin, Y. and Pennington, M. (2000) 'Public participation and local environmental planning: the collective action problem and the potential of social capital', *Local Environment*, 5, 153-169.

Saarman, E. and Carr, M. (2013) 'The California Marine Life Protection Act MPA network: a governance analysis', *Marine Policy*, 41, 41-49.

Sala, E., Aburto-Oropeza, O., Parades, G., Parra, I., Barrera, J.C. and Dayton, P.K. (2002) 'A general model for designing networks of marine reserves', *Science*, 298, 1991-1993.

Salomon, A.K., Shears, N.T., Langlois, T.J., Russell, C. and Babcock, R.C. (2008) 'Cascading effects of fishing can alter carbon flow through a temperate coastal ecosystem', *Ecological Applications*, 18, 1874-1887.

Sandel, M.J. (2012) *What Money Can't Buy: the moral limits of markets*, Penguin, London.

Sandin, S.A., Smith, J.E., Demartini, E.E., Dinsdale, E.A., Donner, S.D., Friedlander, A.M., Konotchick, T., Malay, M., Maragos, J.E., Obura, D., Pantos, O., Paulay, G., Richie, M., Rohwer, F., Schroeder, R.E., Walsh, S., Jackson, J.B.C., Knowlton, N. and Sala, E. (2008) 'Baselines and degradation of coral reefs in the Northern Line Islands', *PLoS ONE*, 3, 2, e1548.

Sands, P. (2003) *Principles of International Environmental Law* (2nd edition), Cambridge University Press, Cambridge, UK.

Saunders, F., Mohammed, S.M., Jiddawi, N. and Sjöling, S. (2008) 'An examination of governance arrangements at Kisakasaka Mangrove Reserve in Zanzibar', *Environmental Management*, 41, 663–675.

Schulman, M.D. and Anderson, C. (1999) 'The Dark Side of the Force: a case study of restructuring social capital', *Rural Sociology*, 64, 351–372.

Sciberras, M., Jenkins, S.R., Mant, R., Kaiser, M.J., Hawkins, S.J. and Pullin, A. (2014) 'Evaluating the relative conservation value of fully and partially protected marine areas' *Fish and Fisheries*, DOI:10.1111/faf.12044.

Shears, N.T. and Babcock, R.C. (2002) 'Marine reserves demonstrate top-down control of community structure on temperate reefs', *Ocealogia*, 132, 131–142.

—(2003) 'Continuing trophic cascade effects after 25 years of no-take marine reserve protection', *Marine Ecology-Progress Series*, 246, 1–16.

Shipp, R.L. (2003) 'A perspective on marine reserves as a fishery management tool', *Fisheries*, 28, 12, 10–21.

—(2004) 'Harvest benefits: marine reserves or traditional fishery management tools', *American Fisheries Society Symposium*, 42, 125–131.

Silliman, R.P. (1975) 'Selective and unselective exploitation of experimental populations of *Tilapia mossambica*', *Fishery Bulletin*, 78, 3, 495–507.

Smith, M.D., Roheim, C.A., Crowder, L.B., Halpern, B.S., Turnipseed, M., Anderson, J.L., Asche, F., Bourillón, L., Guttormsen, A.G., Khan, A., Liguori, L.A., McNevin, A., O'Connor, M.I., Squires, D., Tyedmers, P., Brownstein, C., Carden, K., Klinger, D.H., Sagarin, R. and Selkoe, K.A. (2010) 'Sustainability and global seafood', *Science*, 327, 784–786.

Sobel, J. and Dahlgren, C. (eds) (2004) *Marine Reserves: a guide to science, design and use*, Island Press, Washington DC.

Spotila, J.R., Reina, R.D., Steyermark, A.C., Plotkin, P.T. and Paladini, F.V. (2000) 'Pacific leatherback turtles face extinction', *Nature*, 405, 529–530.

Steele, J.F. (1991) 'Marine ecosystem dynamics: comparisons of scale', *Ecological Research*, 6, 175–183.

Stein, H. (1994) *Adam Smith did not wear an Adam Smith necktie*. Wall Street Journal, April 6, 1994. Commentary on misinterpretations of Smith, A. (1776) *The Wealth of Nations*.

Steins, N.A. and Edwards, V.M. (1999) 'Synthesis: platforms for collective action in multiple-use common-pool resources', *Agriculture and Human Values*, 16, 309–315.

Steneck, R.S., Graham, M.H., Bourque, B.J., Corbett, D., Erlandson, J.M., Estes, J.A. and Tegner, M.J. (2002) 'Kelp forest ecosystems: biodiversity, stability, resilience and future', *Environmental Conservation*, 29, 436–459.

Stern, P.C., Dietz, T., Dolšak, N., Ostrom, E. and Stonich, S. (2002) 'Knowledge and questions after 15 years of research', pp. 445–489 in E. Ostrom, T. Dietz, N. Dolšak, P.C. Stern, S. Sonich and E.U. Weber (eds) *The Drama of the Commons*, National Academy Press, Washington DC.

Stobart, B., Warwick, R., González, C., Mallol, S., Díaz, D., Reñones, O. and Goñi, R. (2009) 'Long term and spillover effects of a marine protected area on an exploited fish community', *Marine Ecology Progress Series*, 384, 47–60.

Stokstad, E. (2009) 'Détente in the fisheries war', *Science*, 324, 170–171.

Swan, J. and Gréboval, D. (2005) 'Part 1: Report of the International Workshop on the implementation of international fisheries instruments and factors of unsustainability and overexploitation', pp. 3–16 in J. Swan and D. Gréboval (eds) *Overcoming Factors of Unsustainability and Overexploitation in Fisheries: selected papers on issues and approaches*, FAO Fisheries Report No. 782, FAO, Rome.

Symes, D. (2000) 'Use rights and social obligations: questions of responsibility and governance', pp. 276–283 in R. Shotton (ed.) *Use of Property Rights in Fisheries Management*, FAO Fisheries Technical Paper 404/1, FAO, Rome.

TNC/UNEP-WCMC (2012) *Policy Brief – Aichi Target 11: reshaping the global agenda for MPAs*. The Nature Conservancy, Arlington, USA/United Nations Environment Programme-World Conservation Monitoring Centre, Cambridge, UK.

Taylor, E., Baine, M., Killmer, A. and Howard, M. (2013) 'Seaflower Marine Protected Area San Andres Archipelago, Colombia', *Marine Policy*, 41, 57–64.

Terborgh, J. (1999) *Requiem for Nature*, Island Press, Washington DC.

Thrush, S.F. and Dayton, P.K. (2002) 'Disturbance to marine benthic habitats by trawling and dredging: implications for marine biodiversity', *Annual Review of Ecology and Systematics*, 33, 449–473.

Thrush, S.F., Gray, J.S., Hewitt, J.E. and Ugland, K.I. (2006) 'Predicting the effects of habitat homogenization on marine biodiversity', *Ecological Applications*, 16, 1636–1642.

Thurstan, R.H., Brockington, S. and Roberts, C.M. (2010) 'The effects of 118 years of industrial fishing on UK bottom trawl fisheries', *Nature Communications*, 1, Article 15, DOI:10.1038/ncomms1013.

Thurstan, R.H., Hawkins, J.P., Neves, L. and Roberts, C.M. (2012) 'Are marine reserves and non-consumptive activities compatible? A global analysis of marine reserve regulations', *Marine Policy*, 36, 1096–1104.

Toropova, C., Meliane, I., Laffoley, D., Matthews, E. and Spalding, M. (eds) (2010) *Global Ocean Protection: Present Status and Future Possibilities*. IUCN, Gland, Switzerland, The Nature Conservancy, Arlington, USA, UNEP-WCMC, Cambridge, UK, UNEP, Nairobi, Kenya, UNU-IAS, Tokyo, Japan, Agence des aires marines protégées, Brest, France.

Tushman, M.L. (1977) 'Special boundary roles in the innovation process', *Administrative Science Quarterly*, 22, 587–605.

UNCLOS (1982) *United Nations Convention on the Law of the Sea*, http://www.un.org/depts/los/convention_agreements/texts/unclos/closindx.htm accessed 4 February 2012; further reading: Churchill, R.R. and Lowe, A.V. (1999) *The Law of the Sea*, 3rd edition, Melland Schill Studies in International Law, Manchester University Press, Manchester, England.

UNEP (1995) *Global Biodiversity Assessment*, Cambridge University Press, Cambridge, UK.

——(2006) *Marine and Coastal Ecosystems and Human Wellbeing: a synthesis report based on the findings of the Millennium Ecosystem Assessment*. United Nations Environment Programme, Nairobi.

——(2010) *Global Synthesis – a report from the Regional Seas Conventions and Action Plans for the Marine Biodiversity Assessment and Outlook series*. United Nations Environment Programme, Nairobi.

Ulrich, R.S. (1993) 'Biophilia, biophobia and natural landscapes', pp. 73–137 in S.R. Kellert and E.O. Wilson (eds) *The Biophilia Hypothesis*, Island Press, Washington DC.

Vandeperre, F., Higgins, R.M., Sánchez-Meca, J., Maynou, F., Goñi, R., Martín-Sosa, P., Pérez-Ruzafa, A., Afonso, P., Bertocci, I., Crec'hriou, R., D'Anna, G., Dimech, M., Dorta, C., Esparza, O., Falcón, J.M., Forcada, A., Guala, I., Direach, L.L., Marcos, C., Ojeda-Martínez, C., Pipitone, C., Schembri, P.J., Stelzenmüller, V., Stobart, B. and Santos, R.S.

(2011) 'Effects of no-take area size and age of marine protected areas on fisheries yields: a meta-analytical approach', *Fish and Fisheries*, 12, 412-426.

Veitch, L., Dulvy, N.K., Koldeway, H., Lieberman, S., Pauly, D., Roberts, C.M., Rogers, A.D. and Baillie, J.E.M. (2012) 'Avoiding empty ocean commitments at Rio+20', *Science*, 336, 1383-1385.

Verhulst, S., Oosterbeek, K., Rutten, A.L. and Ens, B.J. (2004) 'Shellfish fishery severely reduces condition and survival of Oystercatchers despite creation of large marine protected areas', *Ecology and Society*, 9, 1, article 17.

WRI (2011) *Reefs at Risk Revisited*. World Resources Institute, Washington, DC.

Walker, B.H., Gunderson, L.H., Kinzig, A.P., Folke, C., Carpenter, S.R. and Schultz, L. (2006) 'A handful of heuristics and some propositions for understanding resilience in social-ecological systems', *Ecology and Society*, 11, 1, article 13.

Walker, B.H., Holling, C., Carpenter, S.R. and Kinzig, A.P. (2004) 'Resilience, adaptability and transformability in social-ecological systems', *Ecology and Society*, 9, 2, article 5.

Walters, B.B. (2004) 'Local management of mangrove forests in the Philippines: successful conservation or efficient resource exploitation?', *Human Ecology*, 32, 177-195.

Walters, C.J. and Hilborn, R. (1978) 'Ecological optimization and adaptive management', *Annual Review of Ecology and Systematics*, 9, 157-188.

Walther, G.-R., Post, E., Convey, P., Menzel, A., Parmesan, C., Beebee, T.J.C., Fromentin, J.-M., Hoegh-Guldberg, O. and Bairlein, F. (2002) 'Ecological responses to recent climate change', *Nature*, 416, 389-395.

Watling, L. and Norse, E.A. (1998) 'Disturbance of the seabed by mobile fishing gear: a comparison to forest clearcutting', *Conservation Biology*, 12, 1180-1197.

Watson, R.A., Cheung, W.W.L., Anticamara, J.A., Sumaila, R.A., Zeller, D. and Pauly, D. (2012) 'Global marine yield halved as fishing intensity redoubles', *Fish and Fisheries*, 14, 493-503.

Weisman, W. and McCay, B. (2011) 'Isla Natividad Marine Protected Area – governance analysis', pp. 156-163 in P.J.S. Jones, W. Qiu and E.M. De Santo (eds) *Governing Marine Protected Areas: getting the balance right – Volume 2*, UNEP, Nairobi.

Western, D. and Wright, R.M. (1994) 'The background to community-based conservation', pp. 1-14 in D. Western, R.M. Wright and S.C. Strum (eds) *Natural Connections: perspectives in community-based conservation*, Island Press, Washington DC.

Whitley, D. (2008) *The Idea of Nature in Disney Animation*. Ashgate, Surrey, England.

Wilson, J.A. (2006) 'Matching social and ecological systems in complex ocean fisheries', *Ecology and Society*, 11, 1, article 9.

Wood, A., Stedman-Edwards, P. and Mang, J. (2001) *The Root Causes of Biodiversity Loss*, Earthscan, London.

Wood, L.J., Fish, L., Laughren, J. and Pauly, D. (2008) 'Assessing progress towards global marine protection targets: shortfalls in information and action', *Oryx*, 42, 340-351.

Worm, B., Barbier, E.B., Beaumont, N., Duffy, J.E., Folke, C., Halpern, B.S., Jackson, J.B.C., Lotze, H.K., Micheli, F., Palumbi, S.R., Sala, E., Selkoe, K.S., Stachowicz, J.J. and Watson, R. (2006) 'Impacts of biodiversity loss on ocean ecosystem services', *Science*, 314, 787-790.

Worm, B., Davis, B., Kettemer, L., Ward-Paige, C.L., Chapman, D., Heithaus, M.R., Kessel, S.T. and Gruber, S.H. (2013) 'Global catches, exploitation rates, and rebuilding options for sharks', *Marine Policy*, 40, 194-204.

Worm, B., Hilborn, R., Baum, J.K., Branch, T.A., Collie, J.S., Costello, C., Fogarty, M.J., Fulton, E.A., Hutchings, J.A., Jennings, S., Jensen, O.P., Lotze, H.K., Mace, P.M.,

McClanahan, T.R., Minto, C., Palumbi, S.R., Parma, A., Ricard, D., Rosenberg, A.A., Watson, R. and Zeller, D. (2009) 'Rebuilding global fisheries', *Science*, 325, 578–585.

Yardley, W. and Olsen, E. (2011) 'With powerboat and forklift, a sacred whale hunt endures', *New York Times*, 16 October, 2011, http://www.nytimes.com/2011/10/17/us/in-sacred-whale-hunt-eskimos-use-modern-tools.html, accessed 10 July 2012.

Young, O.R. (2010) 'Institutional dynamics: resilience, vulnerability and adaptation in environmental and resource regimes', *Global Environmental Change*, 20, 378–385.

Young, O.R., Osherenko, G., Ekstrom, J., Crowder, L.B., Ogden, J., Wilson, J.A., Day, J.C., Douvere, F., Ehler, C.N., McLeod, K.L., Benjamin, S., Halpern, B.S. and Peach, R. (2007) 'Solving the crisis in ocean governance: place-based management of marine ecosystems', *Environment*, 49, 4, 20–32.

Žydelis, R., Small, C. and French, G. (2013) 'The incidental catch of seabirds in gillnet fisheries: a global review', *Biological Conservation*, 162, 76–88.

Glossary

Italicised words are defined in another entry in this glossary

Actors   People involved in a given MPA *governance* initiative, including local *users*, representatives of governmental and non-governmental organisations, etc.

Basic conflicts   Conflicts based on differences between utilitarian values, focused on exploiting marine resources, and ecocentric-preservationist values, focused on conserving *ecosystem health* and setting areas aside from direct human uses, often revealed in the context of MPAs between those *actors* focused more on utilitarian objectives and those focused more on conservationist objectives.

Biodiversity   The diversity of different ecosystems, habitats and species, including genetic diversity amongst different populations of a given species.

Decentralisation   The transfer of authority from central government to lower-level government levels, quasi-independent government organisations, NGOs or the private sector, degrees and forms of autonomy ranging from deconcentration, to delegation, to devolution (Rondinelli, 2000; Oxhorn, 2004).

Driving forces   The factors that can promote activities that can undermine effectiveness, such as increasing human populations, both from local population growth and inward migration, increasing demands from globalised fish and tourism markets, and the increasing aspirations of people to improve their living standards beyond subsistence livelihoods.

Ecosystem health   A measure of the structural and functional integrity, biological diversity and *resilience* of marine ecosystems coupled with their capacity to provide sustainable flows of *ecosystem services*.

Ecosystem services   'The direct and indirect use benefits people obtain from ecosystems' (Beaumont et al., 2007), such as food provision, nutrient recycling, climate regulation and shoreline protection.

Effectiveness   The degree to which an MPA's objectives have been achieved and related obligations fulfilled, through the control of impacts, involving restrictions on activities to which an MPA's habitats and species are sensitive.

Equity   The fair distribution of costs (related to use restrictions) and benefits (related to the achievement of objectives) arising from MPAs, including recognition of the importance of local ways of life.

Governance   The involvement of a wide range of *institutions* and *actors* in the production of policy outcomes ... involving coordination through networks and partnerships' (Johnston et al., 2000, 317).
   or
   Steering human behaviour through combinations of state, market and civil society approaches in order to achieve strategic objectives.

Incentives   A particular *institution* that is instrumentally designed in relation to a MPA to encourage actors to choose to behave in a manner that provides for certain strategic policy outcomes, particularly biodiversity conservation and restoration objectives, to be achieved.

Institutions   Prescriptions that humans use to organise all forms of repetitive and structured interactions, including those within families, neighbourhoods, markets, firms, sports leagues, churches, private associations, and governments at all scales (Ostrom, 1995, 3).

Management   The day-to-day control of users and their activities, including related technical and administrative approaches.

No-take MPAs   Marine areas designated for the conservation and restoration of their ecosystems, where all fishing activities are permanently banned, as are all other activities that involve the removal of living and non-living resources, e.g. recreational angling, shellfish collection, sand extraction.

Partially protected MPAs   Marine areas designated for the conservation and restoration of particular habitats and/or species, which provide for some activities that are compatible with such objectives on the basis that they do not significantly impact the particular habitats or species, such as recreational angling and commercial fishing with static gears (traps, pots, set nets, etc.) and pelagic trawls (towed through the water column, not across the seabed).

Resilience   A measure of the persistence of systems and of their ability to absorb change and disturbance and still maintain the same relationships between populations or state variables (Holling, 1973).

Social capital   A measure of the degree to which *actors* reach and implement decisions together through their professional and social networks, placing trust in one other, and having confidence that their cooperation with measures to achieve agreed collective objectives will be reciprocated by other actors.

Stakeholders   People who have a stake in a given MPA as they are direct or indirect *users* and thereby benefit from *ecosystem services*. This is generally confined to *users*, but some definitions are more akin to *actors* in that they include representatives of state organisations, NGOs, etc. whilst others include members of wider society who may gain more distant indirect benefits, sometimes even extending to future generations. Due to the ambiguity of this term, it is only used where appropriate to the case study or other context.

Users   People who use a MPA on a direct basis, by extracting natural resources, or on an indirect basis, through non-extractive recreational activities, aesthetic appreciation, etc. For the purposes of this study, users are confined to those who live in the locality of the MPA or who often visit it for direct and indirect uses, i.e. it excludes members of wider society who may gain more distant indirect benefits. Representatives of state organisations are not considered as users.

Glossary of acronyms

BINGO	Big international NGO
CFMAs	Conventional fisheries management approaches
CFP	Common Fisheries Policy (EC)
CPRs	Common-pool resources
CPUE	Catch-per-unit-effort
EC	European Commission
EU	European Union
GBRMP	Great Barrier Reef Marine Park
GNPS	Galápagos National Park Service
ITQ	Individual transferable quota
IUU	Illegal, unreported and unregulated (fishing)
LEDC	Less economically developed country
LMMA	Locally managed marine area
MEDC	More economically developed country
MLPA	Marine Life Protection Act (California)
MPA	Marine Protected Area
MPAG	MPA governance
MSP	Marine spatial planning
MSY	Maximum sustainable yield
NGO	Non-governmental organisation
NMS	National Marine Sanctuary (US)
NTZ	No-take zone
PES	Payments for ecosystem services
RLFF	Resources Legacy Fund Foundation
TURFs	Territorial user rights in fisheries
UK	United Kingdom
UNCLOS	United Nations Convention on the Law of the Sea
US	United States

Index

Italicised entries refer to titles of documents. Page numbers containing 't' refer to tabulated information, those containing 'f' refer to figures, and those containing 'b' refer to boxes.

Lightning Source UK Ltd.
Milton Keynes UK
UKOW06f1849160916

283180UK00013B/321/P